*When Physics Became King*

# WHEN PHYSICS
## BECAME KING

Iwan Rhys Morus

*The University of Chicago Press*
*Chicago & London*

Iwan Rhys Morus is a lecturer in the Department of History and Welsh History at the University of Wales, Aberystwyth, and coauthor of *Making Modern Science: A Historical Survey*, also published by the University of Chicago Press.

The University of Chicago Press, Chicago 60637
The University of Chicago Press, Ltd., London
© 2005 by The University of Chicago
All rights reserved. Published 2005
Printed in the United States of America

14 13 12 11 10 09 08 07 06 05     1 2 3 4 5

ISBN: 0-226-54201-7    (cloth)
ISBN: 0-226-54202-5    (paper)

Library of Congress Cataloging-in-Publication Data

Morus, Iwan Rhys, 1964–
    When physics became king / Iwan Rhys Morus.
        p. cm.
    Includes bibliographical references and index.
    ISBN 0-226-54201-7 (alk. paper) – ISBN 0-226-54202-5 (pbk. : alk.paper)
1. Physics–Europe–History–19th century.    I. Title.

QC9.E89M67 2005
530′.094–dc22                                        2004015207

⊗ The paper used in this publication meets the minimum requirements of the American National Standard for Information Sciences—Permanence of Paper for Printed Library Materials, ANSI Z39.48-1992.

*To the memory of my darling wife, Bridgheen,*
*who made it all worthwhile.*

# Contents

# Illustrations

# Acknowledgments

This book has been a long time in the making. It is an attempt to provide an engaging and accessible cultural history of nineteenth-century physics that brings together the significant developments in our understanding of that field and its cultural connections over the last few decades. I have enjoyed writing it immensely and, while doing so, have managed to surprise myself both by how much and by how little I know about the subject that is meant to be my main area of historical expertise.

As will become readily apparent, much of what I have written here is heavily dependent on other historians. I am deeply indebted to them and hope that they do not feel after reading this book that I have taken too many liberties with their writings. The list of debts could go on for ever, but I would like to mention in particular Will Ashworth, David Cahan, Michael Crowe, Robert Fox, Graeme Gooday, Bruce Hunt, Richard Noakes, Simon Schaffer, Crosbie Smith, Andy Warwick, and Norton Wise as individuals to whose work I am especially indebted in particular parts of the following pages. Any errors of fact and interpretation are, of course, entirely my own.

More generally, I have learned a great deal over the years in conversations with friends and colleagues about the history of science. I would like to thank them all, especially my former teachers and fellow students at the Department of History and Philosophy of Science at the University of Cambridge, but in particular Rob Iliffe and Andy Warwick, with whom I was lucky enough to be a contemporary while all kinds of exciting new developments were taking place in the history of science, as well as Simon Schaffer, the source for so many of those developments. I would also like to thank my colleagues at the School of Anthropological Studies at Queen's University Belfast for all their support over the years. The late Susan Abrams at the University of Chicago Press was a constant source of enthusiasm and encouragement. I will always be grateful for her faith in me. Christie Henry at the press continued the enthusiasm and encouragement through some very difficult times. I would also like to thank my copyeditor Michael Koplow for his diligence and hard work.

Finally, I thank my late wife Bridgheen, who was always there for me.

# 1

## Queen of the Sciences

At the entrance to Belfast's Botanic Gardens, across Queen's University's main campus from my office, stands a statue of the physicist William Thomson, Lord Kelvin (figure 1.1). Born in Belfast—where his father taught mathematics at the Academical Institution—in 1824, Thomson's career as a man of science spanned the nineteenth century that forms the focus for this book. The statue is worth a closer look. As he gazes out towards University Road, Thomson holds an open book in his hands. The pages are inscribed with illustrations of his vortex model of the atom—the foundation to one of Thomson's many claims to a place in the physicists' hall of fame. He is leaning backwards slightly against a short pillar that turns out to be a magnetic compass on its stand—a key patented invention of his that played a vital role in the late nineteenth-century shipping industry that powered the economies of both his native Belfast and his adopted city of Glasgow. The combination of civic pride, philosophical insight, and industrial prowess represented by this statue says a great deal about nineteenth-century physics and what it came to stand for. It shows what mattered about physics in late nineteenth-century culture. This book follows the story of physics throughout that century, showing how a science that barely existed in 1800 came to be regarded a hundred years later as the ultimate key to unlocking nature's secrets.

1.1 The statue of William Thomson, Lord Kelvin, that stands at the
entrance to the Botanic Gardens in Belfast. The book that Kelvin is
holding in his hand contains diagrams of his vortex model of the
atom. He is leaning against his patented magnetic compass.

So why does the history of physics matter? One way of answering
the question is simply to point to the central role physics and physicists
have played and continue to play in our culture throughout the twentieth
century and into the twenty-first. One area at least where we still seem
to be largely in agreement with our late Victorian predecessors is on the
issue of physics' preeminence. We take it for granted that physicists are

the best people to find out about the nature of the universe, and we look to their laboratories to keep producing the innovations that power our economies and satisfy our ever-increasing appetites for more. We look to them as well for solutions to the looming environmental catastrophe that two centuries of industrial expansion seems set to deliver. On the whole, but with increasing uncertainty, we still trust physics and physicists. In many ways, the iconography of Kelvin's statue still rings true today. Great physicists are a source of national pride—living embodiments of a country's genius. The number of Nobel laureates in physics is still cited as an index of a country's international reputation. We admire physicists for their insights into the workings of nature. We gasp at the seemingly endless parade of wonders emanating from their laboratories as we look forward to Star Trek–style warp drives and transportation pads. In other ways, however, the status of physics is less clear than it was only a few decades ago. Arguably at least, molecular biology in the form of the Human Genome Project and its offshoots has replaced it in the public eye as the most visible scientific generator of excitement and progress.

As a historian, I am convinced that the first step towards understanding the central role that physics continues to play in twenty-first-century culture—and the ways in which that role is increasingly under threat—is to understand how that role came about in the first place. Peculiar as it might seem from the modern perspective of people accustomed to turning to physics (and to science and technology more generally) as a source of authority and of answers, it was not always like this. Two hundred years ago there was no discipline called physics, nor was there anybody who called themselves a physicist. For most of the century described in this book even, very few practitioners would have described what they did as physics and even fewer would have described themselves as physicists. Ironically enough, Lord Kelvin is one who would have rejected the labels out of hand. The practice of natural philosophy and the natural philosophers who practiced it had very different cultural roles from the ones their descendants play today. There was nothing inevitable about physics' rise to prominence either. Making physics into the dominant discipline it is today took a great deal of work and effort. It was not just a matter of producing theories that appeared more and more successful at accounting for nature. It was a matter of persuading others that these theories really were true, that physics really was the best way of finding out about how the laws of nature operated. In other words, making physics into the preeminent scientific discipline needed real cultural engagement on the part of its practitioners.

This is why the history of physics that I offer here is an unashamedly cultural history. Again, this might at first appear a little strange. It is central to the view of science—and of physics as the preeminent science— that we hold in our culture that science and culture do not mix. In many ways one of the reasons we think we trust physics so much is because it appears to be free of that cultural taint. Physics simply tells it the way it is. Its results have nothing to do with its practitioners or the kinds of institutions they work in or their cultural status. Even the most superficial history of science tells us that this view is wrong, however. Physicists have always engaged with their culture, as did natural philosophers before them. Indeed, as this history of physics will show, without this kind of engagement there simply would not be any such thing as physics today. Making physics work needed more than great experiments and new insights into the workings of nature. At the very least those great experiments needed laboratories where they could be carried out. That meant persuading others to provide the resources needed to build those laboratories. Similarly, great insights into the workings of nature are of very little use unless others can be persuaded that they are worth listening to. This means that physicists in the past had to establish their own authority and their own claims to people's trust at just the same time as they went about establishing their science.

The long standing perception of an unbridgeable chasm between science and 'culture' is another good reason for insisting that a properly cultural history of physics is essential. Physics is often portrayed—even by some of its greatest promoters—as some kind of alien force. The wild-eyed, disheveled physicist (think Albert Einstein on a bad hair day) is a familiar icon. The image is useful to physicists, serving as it does to underline their otherworldliness, the arcane nature of their practices, and their disconnectedness from mundane affairs. The image is just as useful to physics' detractors and for much the same reasons. That physics is difficult, abstruse, and obsessed by detail seems a good enough reason for relegating it and its practitioners to the margins of modern culture. Ignoring physics like this, however, means ignoring one of the most influential aspects of our culture. Regardless of cozy assumptions by some physicists and their detractors alike that physics has nothing to do with culture, there are in fact very few aspects of our daily lives that are untouched by what physicists do. Understanding this means appreciating that physics is a part of, not apart from, our culture.

A cultural history of physics can also help us to move away from the prevailing view of science as the product of individual great men. When we think of physics' past we typically view it in terms of a succession of individual scientific heroes, each building on the achievements of their predecessor. Most people, including most physicists, probably think of the history of physics (when they think about it at all) in this way. We think of Newton, or Faraday, or Einstein as individuals possessed of some ineffable insight into nature's workings that belongs to them alone. This is what we mean when we describe people like them as geniuses. The picture of physics that emerges from cultural history, however, is very different. Physics from this perspective is a collective enterprise. To understand the way nature works requires collaborative action as much as individual thought. Physics as we know it today is a product of the mass mobilization of material and social resources on an unprecedented scale. To make physics what it is, physicists in the past had to do more than sit in their studies and think. They had to find ways of mobilizing those resources and carving out a cultural niche for themselves and their new discipline. To understand how they did this, we need a cultural history.

The history of science has changed dramatically over the last quarter century. Historians of science up until the 1960s saw their discipline as a handmaiden to philosophy. Philosophers analyzed the scientific method, and historians looked to the past to find examples of the method in action. The history of science was the history of progress. It recorded the gradual accumulation of scientific facts and theoretical insights that led inevitably to our modern understanding of the universe. A new generation of historians borrowed a term from political history to castigate this tradition as "Whig history"—it was the equivalent for science of liberal historians regarding the past in terms of the inevitable rise of liberal democracy and suffered from the same problems in that it judged the actions, achievements, and failings of the past through modern eyes rather than trying to understand them on their own terms. Historians of science in general and of physics in particular now tend to focus their attention on what might be called the microhistory of science instead. They look at particular controversies, the work of particular laboratories or institutions, the consolidation of particular theories. The result is a far richer and more nuanced understanding of physics and how it works than we had before.

Old-style history of science did, however, have one virtue. It had a "big picture." The history of physics as the history of progress could be painted with a broad brush on a large canvas. Historians of physics now are quite rightly suspicious of such big pictures. We know very well that science is far too messy and human an affair for overeasy general-ization. The problem with this is that it leaves us as historians without a broader perspective. The history of physics written as the history of progress from ignorance to enlightenment did at least provide its authors and their readers with a convenient narrative framework from which to hang their portraits. This is something that a cultural history of physics should be able to provide as well, however. It has the potential to develop new overarching themes that can play a role in reversing the current frag-mentation of the historical picture. Looking, for example, at the different ways in which physicists have in the past tried to establish their author-ity as the ultimate arbiters on questions about the natural world and the ways in which this might be related to the kind of knowledge they pro-duced does not commit us to any particular view concerning progress or the scientific method. It does, however, give us a new common thread running through our histories.

This particular history of physics starts around 1800. In many ways the choice of starting point is arbitrary. The period covered by the book is a reflection of the period about which I think I have something sen-sible and interesting to say. It is not, however, entirely arbitrary, and the reasons why it is not so provide another indication of the advantages of looking at physics through the lens of culture. Historians are largely agreed that events taking place in the last quarter or so of the eighteenth and around the beginning of the nineteenth century were truly revolu-tionary in their impact. The American and French Revolutions shattered the old social order. The Revolutionary and Napoleonic Wars in Europe convulsed society. The gathering pace of the Industrial Revolution had major implications for national and international economies and the organization of labor and trade. As the nineteenth century moved on, democracy gradually became less of a dirty word as newly powerful social groups sought to match their rising economic power with equivalent political clout. Some historians of science have labeled this period as the Second Scientific Revolution, with consequences as momentous as those of the first, seventeenth-century one. With appropriate qualifications I agree with them.

Physics in anything resembling the modern sense was born out of the cultural cauldron of late eighteenth- and early nineteenth-century Eu-rope and America. It is no accident that the word *physicist*—in English at least—was first coined in the 1830s. The word was not invented earlier for the simple reason that the kind of person it was intended to describe did not exist more than a few years previously. The equivalent French and German words (*physicien* and *Physiker* respectively), while having a longer pedigree, underwent a similar redefinition during this period too. The kinds of institutions that typify modern physics—the laboratories, university training regimes, and research institutions—have their origins in the nineteenth century. It was during the nineteenth century that physicists forged for themselves the authoritative position as the ultimate legitimate spokespersons for nature that they still to a large degree enjoy today. It was during the nineteenth century that the intimate link between physics and industrial and technological progress that we now recognize was first forged as well. None of this is to suggest that there are not clear continuities between nineteenth-century physics and what went before. As we shall see over the next few pages, nineteenth-century physicists had concerns about the laws of nature, about the best ways of investigating them, and about the position of men of science in society that they certainly shared with their natural philosophical predecessors of the eighteenth, seventeenth, and earlier centuries. All I want to argue here is that the discontinuities were in the end rather more important.

## The Worlds of Natural Philosophy

What can we say then about these continuities and discontinuities? Natural philosophers during the first Scientific Revolution certainly regarded themselves as having brought about a profound and important rupture from the past. They were placing the search for knowledge on a new footing, looking at nature instead of consulting the ancient authorities. Natural philosophers such as Galileo wrote in the vernacular rather than in dusty Latin and poked fun at the staid and unimaginative Schoolmen. He found new audiences for his writings in the urbane courts of Italian city-states like Florence and Venice. New societies devoted to the pursuit of natural knowledge, such as the Accademia del Cimento in Florence, the Académie Royale des Sciences in Paris, and the Royal Society of London for the Promotion of Natural Knowledge, likewise looked to the courts for patronage. Their members saw themselves as far removed from the traditional image of the reclusive Scholar. On the contrary, they were civic-minded gentlemen, committed to making their knowledge both useful and accessible. Universities were largely deemed irrelevant to the making of the New Science. Increasingly, it was outside the cloisters in royal or princely courts and aristocratic households that gentlemanly natural philosophers hoped to make their mark and cultivate patronage.

**1.2** The frontispiece of Francis Bacon's *Instauratio Magna* (1620) showing a ship
setting out between the Pillars of Hercules onto the boundless sea of
knowledge. The Latin motto reads "Many will pass through and knowledge will
be increased."

In many ways Francis Bacon had provided the blueprint for these new natural philosophical institutions with his utopian vision of Solomon's House (figure 1.2). Bacon visualized the search for new knowledge as a systematic and thoroughly organized process, based on the procedures of English common law with which, as a prominent Elizabethan courtier, he was intimately familiar. Central to this vision was the view that natural philosophy was something with which the state should concern itself. Bacon wanted the new philosophy to be at the service of the "commonwealth." At a time of widespread religious and political upheaval in England and in Europe more generally he argued that the production of knowledge had to be the preserve of those best fitted to put it to the proper use. The founders of both the Royal Society of London (established in 1660) and the Académie Royale des Sciences in Paris (established in 1666) certainly had Bacon in mind. They were to be seats of responsible and collaborative learning, far removed from the disputatious and unworldly universities. They were to be governed by rules and protocols to maintain proper decorum in speech and action. The members of the French académie were to be salaried servants of the state (figure 1.3), though its English counterpart never achieved as much by way of state or royal patronage despite its best efforts.

Even the English Crown, however, was perfectly willing to pay for natural philosophy when it suited its interests. When Charles II established the Royal Observatory at Greenwich in 1675 he had more reasons for doing so than a desire to further abstract knowledge of the heavens. Astronomy was widely regarded as the key to better navigation, and better navigation as the key to maritime supremacy. Knowing the night skies in detail could help England's navy know exactly where they were at sea. This was the motivation behind the establishment of a large financial award for solving the problem of longitude as well. Throughout the eighteenth century this was the Holy Grail of observational astronomy. To further the quest for useful astronomical knowledge the English Admiralty funded expeditions to observe eclipses and other rare celestial phenomena such as the Transit of Venus in 1761 and again in 1769 as part of the voyages of Captain Cook. The French government similarly financed expeditions like this, including one famous expedition to Lapland in 1736 to make measurements leading to more accurate calculations of the shape of the Earth. Natural philosophy was regarded as a tool that could be used to expand and defend the power of the state.

Natural philosophers were increasingly keen to find other patrons as well. In eighteenth-century Amsterdam, London, Paris, and Philadelphia

1.3 An imaginary visit to the Académie Royale des Sciences by its patron, Louis XIV, the king of France. The scene makes explicit the intimate connections between the new scientific institution and the French monarch.

natural philosophers mingled with wheeler-dealer entrepreneurs in coffeehouses as they tried to market their science. They offered steam engines to pump water from flooded mines, or schemes of improving agricultural methods or of building canals. Natural philosophy could be made useful and lucrative. Popular lecturers in coffeehouses and polite salons offered spectacular demonstrations of their ability to control the powers of nature as part of their sales patter. Electricians such as the Dutchman Martinus van Marum or the Englishman William Cuthbertson could build machines that seemed to mimic lightning and also offer schemes that might protect vulnerable buildings from fires caused by lightning

strikes. The American Benjamin Franklin had more than just philosophical fame in mind with his invention of the lightning rod. Philosophical
showmen were part of the entrepreneurial culture of eighteenth-century
Europe and America. Their improvement schemes fitted in well with
the prevailing expansionist ethos. Enlightenment commentators saw the
rational knowledge that natural philosophy offered as the solution to
humankind's miseries and as a way of remodeling society. Natural philosophy was becoming an institution—a part of a common public civic
culture.

The audience for natural philosophy was increasingly drawn from the
rising urban middle class. These were the men (and women) who flocked
to public lectures for entertainment and edification. In England, many
in the provincial middle classes came from Dissenting backgrounds—
outside the Anglican Church and therefore excluded from access to political power. They turned to natural philosophy as a channel for their
cultural aspirations. It could be a way of justifying their claims to political
power as well. Members of the Lunar Society such as the radical Joseph
Priestley, the manufacturer Josiah Wedgwood, and the physician Erasmus
Darwin reckoned that their natural philosophy provided a model for
the reform of society. By the end of the eighteenth century, literary and
philosophical societies were spreading the taste for science to the aspir-
ing provincial middle classes. Natural philosophy was seen as a model
for overturning the social order in France as well. Friend and foe alike
regarded new natural powers like galvanism or oxygen as revolutionary
spirits. Back in England the identification was so complete that Joseph
Priestley fled to newly independent America after his Birmingham laboratory was looted and burned by a monarchist mob. Such leaders of
the fledgling United States as Benjamin Franklin and Thomas Jefferson
agreed with his view of natural philosophy as the model for an enlight-
ened social order.

Throughout the seventeenth and eighteenth centuries, natural
philosophers accommodated themselves to constantly changing social
and political circumstances. Just as they worked to fashion new ways
of understanding the natural world, they had to work to find a place
for themselves in society as well. Knowledge could very easily be seen
as a dangerous thing. The role of natural philosophy in sustaining (and
subverting) the existing political and religious order was under constant
scrutiny. Seventeenth-century natural philosophers had to find ways of
distinguishing themselves and the knowledge they offered from dangerous religious enthusiasts. Likewise, their eighteenth-century descendants

had to guard themselves from accusations of stagecraft and charlatanism from their conservative critics, particularly as natural philosophy and political radicalism came to be more closely associated. Natural philosophers had constantly to reinvent themselves and their science in order to try to find secure social niches for themselves and their practices. Philosophical arguments about the nature of knowledge, the best way of investigating the natural world, and the kind of people best fitted to carry out such investigations merged easily with straightforwardly political discussions of the nature of political power and its distribution throughout society. Charting a safe route through these waters demanded constant circumspection.

Mathematics was one route that seemed to its supporters to offer a sure means of establishing their knowledge of nature on a secure basis. Mathematics had the virtue that its procedures seemed to guarantee certainty in its conclusions. So long as the axioms from which the mathematician proceeded to theorize were certain, then the conclusions he arrived at were certain as well. Galileo, who started out as a professor of mathematics, sought to elevate mathematics (and himself) to the high status of philosophy and bring the solidity of its arguments to bear on questions about nature. The problem he faced was that to his critics mathematics seemed to deal only with trivial things like numbers and quantities, rather than delving into the real essence of things as a true natural philosophy should. Even the illustrious Isaac Newton and his *Philosophiae Naturalis Principia Mathematica* (*Mathematical Principles of Natural Philosophy*) were criticized for failing to account properly for the true causes of the phenomena. Newton's *Principia* was nevertheless widely celebrated as epitomizing the way forwards for natural philosophy and its author hailed as the model man of science. Very few of even its most ardent supporters either among his contemporaries or their later eighteenth-century audience could actually follow the details of its mathematical arguments, however, let alone reproduce them. Mathematics was a science for the elite few, not the common crowd.

Experiment was another candidate that natural philosophers looked to as a way of placing the New Science on a more secure footing. They would put nature to the test and wrest its secrets from it. The proper foundation of knowledge according to this view was detailed and disciplined observation. By looking in detail at the operations of nature, by carrying out experiments that systematically set out to investigate its phenomena under different circumstances, natural philosophers hoped to be able to arrive at fundamental laws. Crucially, experiment was conceived as a

collaborative activity. This was one reason mid-seventeenth-century natural philosophical investigators were so keen to combine in scientific
societies like the Royal Society. Groups of like-minded natural philosophers could meet to discuss their experimental findings, share their results, and combine their knowledge. Experimental philosophy was meant
to be open to the participation of all open-minded gentlemen inclined
towards curiosity who were prepared to investigate diligently and with
due skepticism and diffidence. Its critics, of course, pointed out that it
was no such thing. In practice, it was only the Royal Society's own fellowship that had access to its experimental results. Outspoken opponents of
the experimental method like the irascible English philosopher, Thomas
Hobbes, never became fellows of the Royal Society.

Experimental philosophy depended in particular on the trustworthiness of its practitioners. Audiences had to be convinced of their veracity—
that they were accurately relating the details of things that they had
really seen and experienced. This is one reason why codes of civility
and gentlemanly behavior were held to be crucial to natural philosophy.
Gentlemen's words could be trusted (or so it was commonly held) and
natural philosophers sought to buy into that privileged status by insisting that practitioners should be gentlemen too. Women, tradesmen, and
artisans were disbarred by default from the company of natural philosophers. Their word could not be trusted. Behind the scenes, of course,
artisans did a great deal of the actual work of natural philosophy. It was
Robert Boyle's servants, not himself, who worked his air-pumps. Their
testimony was not deemed to be required when describing the results of
the experiments, however. The word of aristocrats and eminent fellows
of the Royal Society was more trustworthy. Crossing the boundary from
being a servant to being a natural philosopher in one's own right was
difficult, as Robert Hooke, Boyle's assistant who became curator of experiments at the Royal Society, found throughout his precarious career.
Throughout the seventeenth and eighteenth centuries (and as we shall
see, throughout the nineteenth century as well) natural philosophers had
to find ways of convincing sometimes skeptical audiences that they could
indeed be trusted as spokespersons for nature.

Critics could occasionally be scathing of natural philosophers' pretensions to knowledge. Jonathan Swift lambasted scientific societies'
aspirations mercilessly in Gulliver's Travels. On Gulliver's third voyage
he found himself visiting the flying island of Laputa. The island's inhabitants were so distracted by questions of natural philosophy that they
were accompanied everywhere they went by a servant whose task it was

to hit them about the ears with a balloon whenever they were spoken to. Without that reminder they would be too much lost in speculation to notice. He then visited the city of Lagado, which boasted a scientific academy whose members were continually producing useless inventions. Rather than speaking to each other, members of the Academy carried objects around with them instead of using words. This was the antithesis of the image that the civic-minded fellows of the Royal Society or the Académie Royale des Sciences wanted to convey about natural philosophy and their pursuit of it. Conservative critics regarded natural philosophers with suspicion as possible subverters of the social order. They saw little difference between a public lecture or experimental demonstration of the latest electrical or magnetic wonder and the trickery of theatrical performance. Natural philosophers claiming superior knowledge of the operations of nature could be dismissed as mere hucksters out to gull the public or seduce them into revolution.

The politics of natural philosophy was certainly high on its critics' agenda. The same could be said of its promoters as well. Seventeenth- and eighteenth-century natural philosophers were acutely aware of the politics of knowledge. They knew that what was known about nature and who produced the knowledge had important political consequences. This was one reason why the Royal Society's gentlemen or the Académie Royale des Sciences' state-salaried *savants* were so anxious to keep the production of knowledge under their own careful eyes. They needed to be sure that only the suitably qualified were recognized as true men of science. Everyone agreed that the organization of knowledge had political resonances. This was because the social order of things was widely held to mirror the natural order. Looking at the way in which nature was ordered and the laws that governed the operations of nature was a way of finding out about the proper order of society as well. Natural philosophy could therefore either provide a powerful and decisive endorsement of the status quo or produce a recipe for political subversion. Conservative supporters of the *ancien régime* could get distinctly twitchy when radical natural philosophers turned to nature as a way of exposing the deficiencies of the prevailing political order of things.

Clockwork was an increasingly common metaphor for the operations of the Universe. All the parts of a clock worked in harmony to produce the final motion. This was how some natural philosophers visualized the workings of the Universe too. All the parts were in harmony and worked in unison to produce the movements of the Earth and the planets. This had the major advantage of implying the existence of a celestial clockmaker as well. If the Universe were a piece of complex mechanism like a clock, then it must have had a Creator too. The French natural philosopher René Descartes was one of those who took the analogy furthest, suggesting that even the human body should be regarded as a piece of clockwork mechanism acting automatically, though he stipulated that the governing soul was separate from the body. One reason that the clockwork metaphor worked well was that it seemed to provide a good account of the workings of society too. Just like the natural Universe, the social world could be regarded as a complex clockwork mechanism. As long as everyone knew their proper place in society, everything worked smoothly. It was only when something went wrong with the mechanism—when unruly subjects forgot their proper place in the natural order of things—that social order broke down.

The new heliocentric view of the Universe produced by Copernicus fitted in neatly with this world picture. According to this account, the Sun, not the Earth, was at the center of the Universe; all the planets including the Earth revolved around it. The new picture was just as conducive to the clockwork metaphor as the old Earth-centered view, in which the Sun and the other planets revolved around the Earth. It was easily translated into social terms as well. The Sun was the king around which his subjects—the planets—revolved in harmony. It was an ideal metaphor for a time of increasing royal absolutism throughout Europe. Isaac Newton's view of the Universe differed subtly from the clockwork model, however. He argued that God was always immanent and present in nature, continually sustaining the operations of natural laws. This was in opposition to the views of Descartes or the German philosopher Gottfried Wilhelm Leibniz, who argued that a perfect God would have created a perfect mechanism that needed no sustaining. This too had implications for the social order. On the one hand, God's immanence in the world served to support the prevailing social order. At the same time it was an antidote to monarchical absolutism, suggesting that the monarch as much as anyone else be subject to the dictates and limitations of law.

There were other ways of reading the social implications of Newton's theories, however—and these became increasingly popular as the eighteenth century moved on. As far as radical Enlightenment philosophers were concerned, Newton had provided a blueprint for radically reforming the social order. They read Newton as endorsing a natural order that operated by means of a system of divinely ordained checks and balances. The English experimental natural philosopher Joseph Priestley argued that the different powers of nature such as electricity, heat, and magnetism acted to sustain a natural economy. In particular, he suggested that phlogiston—the active power of fire—circulated constantly throughout the natural economy maintaining its balance. For Priestley and others like him, making this natural economy visible was an indictment of the corrupt and despotic English government that failed to live up to the standards of the natural economy. This was the same kind of political ethos as lay behind publications like Thomas Paine's revolutionary *Rights of Man*. The situation was similar in France, where radical political philosophers used natural philosophy to attack the notion of absolute monarchy. When Julien Offrey de la Mettrie argued that humans were no more than soulless machines operating according to the laws of mechanics, he was also arguing against the Church—one of the strongest bulwarks of the French monarchy.

By the end of the eighteenth century, much natural philosophy had a dangerous reputation. For many of its critics it stood accused of being implicated in the political convulsions that swept through Europe during the final decades of the century. In England, espousing natural philosophy—certainly natural philosophy of the kind advocated by men such as Joseph Priestley—was easily seen as tantamount to expressing support for the French Revolution and the Terror that followed. Many radical natural philosophers had indeed been supporters of the revolution, though many also revised their support in the light of subsequent events. Many of the new revolutionary French regime's leaders and ardent supporters were fans or active practitioners of natural philosophy as well. They would have agreed with the critics that their science was indeed implicated in the revolution. Natural philosophy in France, as we shall see, flourished under the revolutionary regime and the Napoleonic rule that followed it. English natural philosophers, on the other hand, were obliged to behave with considerably more circumspection. The destruction of Priestley's laboratory by the mob was a graphic reminder of the perils of experimentation. Some scientific societies even found themselves banned under new Anti-Sedition Laws prohibiting public meetings. Events at the end of the eighteenth century showed clearly just how powerful, but at the same time just how easily marginalized, natural philosophy could be.

## The Birth of Physics

As we shall see, many of these concerns with natural philosophy's social place, the identity and status of its practitioners, and the relationship between the natural and social orders, that had worried men of science from the First Scientific Revolution, still mattered a great deal to their nineteenth-century counterparts. The ways in which those concerns might be expressed and resolved could be very different, however. Looking at the ways in which they tried to resolve these issues will be the major concern of this book. The pages that follow are thematic rather than strictly chronological. That is to say, the story does not start straightforwardly in 1800 and follow a simple time line through to the end of the century. Instead, each chapter will focus on a particular theme in the nineteenth-century history of physics and follow it through the century. In no way is what follows intended to be a comprehensive account of nineteenth-century developments in physics. Such an account would clearly be impossible—it would certainly be so in a book of this size. Instead, I have chosen to focus on a number of episodes and developments that seem to me to capture something important about general trends and particular issues that turned out to be crucial in the development of what we now recognize as the modern scientific discipline of physics.

The story starts in chapter 2 with the rise of mathematical physics in revolutionary France at the beginning of the century and the emergence there of new analytical styles of reasoning about nature. The chapter then follows efforts to develop new institutions and new ways of training mathematical physicists in England and the German states to midcentury and beyond. Chapter 3 looks at various efforts during the century to explore the unity of nature, starting with the Romantic movement at the beginning of the century and culminating in the development of ether physics at its end. Chapters 4 and 5 then focus on the crucial sciences of electricity and heat. We look at how developments in electricity emerged from concerns with showmanship and utility and how the new science of heat responded to the problem of maximizing the efficiency of steam engines in industrial Britain. In chapter 6 we return to electricity, looking at the ways in which it proliferated into a whole range of new forces and powers during the second half of the century, providing ammunition both for and against the increasingly dominant physics of energy and the ether. Chapter 7 looks at developments in astronomy, the consolidation of observational astronomy, and the importation of laboratory science into the study of the heavens. Finally, in chapter 8, we look at the rise of the laboratory and the development of precision measurement as the hallmark of experimental physics before concluding in chapter 9 with a brief survey of physics and its institutions on the eve of the Great War. The story finishes with the First World War for two reasons. In the first place, the war brought the role of the new science of physics in international conflict to prominence for the first time. Physicists and their institutions in both Britain and Germany were involved in the war effort to an unprecedented degree. By the beginning of the twentieth century most of the institutions we identify with physics—the research institutes, university laboratories, and training regimes—were in place. The war made their relationship to the state explicit. At the same time, the decade of the Great War also saw the development of Einstein's theories of special and general relativity along with the beginnings of quantum mechanics. Within very few years the grand consolidation of physics that had emerged around the ether at the end of the nineteenth century, and which many physicists regarded as the end of physics, was in tatters. Ironically, the very aspect of physicists' culture about which they felt most secure proved fallible, while the institutional structures that still seemed very fragile to many of them survived and prospered. If the social convulsions at the nineteenth century's beginning seem a good place to start the story of physics' emergence then, the equally socially tumultuous second decade of the twentieth century seems a good place to draw breath and look back at what emerged.

There are some very clear themes running through all the chapters that follow. The first thing that should be clear is that the main business of physics—the thorough investigation of nature and the effort to understand nature's workings as the outcome of universal physical laws—was crucially dependent on a range of cultural and material resources. As I started this introduction by suggesting, physics does not and cannot take place in a vacuum. In order to be able to successfully investigate the laws of nature, investigators need to be able to get their hands on the right tools for the job. To do physics they need laboratories, they need training in the complexities of mathematical analysis, they need instruments and people with the skills to make those instruments. They also need to share a culture with people who are willing to give them access to these things. One of the threads holding the following narrative together is, therefore, the attempt to find out what kinds of resources nineteenth-century practitioners needed to put together the view of the physical world that they did and how they went about getting those resources.

In particular, the book investigates the cultural and material resources that went into constructing the ether. To many physicists by the end of the nineteenth century—probably to most British physicists—the ether *was* physics. They might not yet have gained a full understanding of the ether's physical properties, but they were as certain of its existence as they were of the existence of the Sun and the planets. They were certainly as sure it existed as modern physicists are of the existence of quarks and neutrinos. So what made them so sure? How did they know that the ether was really there and not just some figment of their imaginations or convenient theoretical construct? In other words, what made their view of the Universe so stable? My answer is that they knew because they could make instruments that could measure its properties, they could manipulate its properties to send information down wires or through space. They also knew because the ether was made up of the same kinds of things, acting in the same kinds of ways, as the visible world around them. It was familiar and made sense in the context of late Victorian industrial culture. In more ways than in a literal sense, late Victorian physicists made the ether from bits of the world around them. In that sense, despite its solidity, it was very much a contingent product of its times.

Another thread running through the book is the importance of institution building in establishing physics as a discipline. Carving out institutional spaces for the new discipline of physics was itself a major cultural accomplishment. There was nothing self-evident at the beginning of the nineteenth century about the idea that universities would become major centers of research into the nature of the physical world, for example. On the contrary, most holders of university sinecures, in England and the German states certainly, would have regarded such a transformation as inimical to their institutions' primary business of teaching established knowledge. Universities were centers of pedagogy, not of the production of new knowledge. In any case, natural philosophy was very far from being the coveted discipline that physics became. It was very much the poor relation in terms of the distribution of university resources and status at the beginning of the nineteenth century. Similarly, the institutional laboratory—now the *sine qua non* of physics research—barely existed in anything like its modern form at the beginning of our period. Neither was there anything like the rigorous training regimes in laboratory-based experimentation and theoretical mathematical analysis that modern physicists undergo as a matter of course. Without these institutional structures there would be no physics as we now recognize it. Again, these were contingent outcomes of determined efforts by nineteenth-century protagonists to carve out institutional spaces for themselves.

Contingency is just as important a lesson to learn for the emergence of physics' institutions as it is for the emergence of physics itself. The institutional history of nineteenth-century physics (and of science more generally) is usually cast as the story of professionalization. Physicists and other scientists during the nineteenth century strove to become professionals—to establish the kinds of institutions that are taken to constitute a profession. There is a great deal to be said for this account. It has the virtue, for example, of making clear the links between developments in the history of science and developments elsewhere in nineteenth-century culture. The problem, however, is that it puts the cart before the horse. Physicists, or at least a large number of people who described themselves as such, were certainly professionals by the end of the century in a way that they were not at its beginnings—they were paid to do physics, for example. Some urge towards becoming professional was not what made them do what they did, however. It would be more accurate to say that the modern sense of what it means to be a professional scientist was the outcome of their activities. The institutions they built and the ways in which they built them came to define for us what kinds of institutions a discipline like physics should have.

Finally, there is a strong sense in which the physicists whose activities I describe in these pages were in the process of making themselves. To establish themselves as the kinds of people who investigated the natural world they had to establish particular sorts of relationship between themselves and their subject matter, their audiences, and the state. What should the relationship between a physicist and the nature they investigate be? What should be the relationship between the physicist and the audience for their researches (and who, for that matter should that audience be)? What should be the role and responsibility of the physicist with regard to the wider community and to the state? None of these questions had clear-cut answers at the beginning of the nineteenth century. It is a moot point whether we have any clearer answers to them now, but trying to find ways of addressing these questions was nevertheless central to the process of making physics both as a discipline and as a reliable way of investigating the physical world. In many ways it boiled down to the question of trust. Physicists had to be the kind of people whom others would trust as reliable witnesses to the way the world really was. They had to persuade their various constituencies that their way of doing it really was the best way.

As I have just indicated, this is a problem that modern physicists need to address as well. They, however, have an advantage over their nineteenth-century predecessors in that they have well-worn institutional paths to follow. Indeed, in many ways that is just what the kinds of institutions whose emergence this book charts are for—they help define who is and who is not a competent and trustworthy practitioner. In the absence of strong and well-supported institutions that defined for practitioners and their audiences what it meant to be a physicist—what kind of training was required and how a competent practitioner should act—individuals had to make their own careers. They had to find ways of defining themselves and what they did in relation to their audiences in such a way as to convince them that they could indeed be trusted to speak for nature. People such as Michael Faraday or Lord Kelvin in Britain, Hermann von Helmholtz in Germany, and Pierre-Simon Laplace in France fashioned themselves in such a way as to become acceptable spokespersons for nature to their respective constituencies. In the absence of anything like a clearly laid-out career path they had to carve one out for themselves. By so doing they helped to define what being a physicist meant.

I hope that this book will provide at least some indication of just how important the history of physics is as a part of understanding the historical development of our own modern industrial and consumer culture. Physics is not, after all, some peculiar esoteric practice out on the fringes of society. It plays a central role in sustaining the way we live today. Thus it is crucial that we try to understand how physics and its institutions operate. Understanding its history is a vital tool in making sense of the present state of physics. History provides us with a salutary reminder of the contingency both of our modern understanding of the physical world and of the institutions where that understanding is forged. We are reminded that there is nothing self-evident or inevitable about the way things are now. Looking at history also provides us with a stark reminder that physics is—irrevocably—a part of culture and a part therefore of our cultural heritage. It is not some alien way of going on. On the contrary, it is part of who we are, informed by and informing the past and present cultures that formed all of us.

# 2

## A Revolutionary Science

When the Parisian crowds stormed the Bastille fortress and prison on 14 July 1789, they set in motion a train of events that revolutionized European political culture. To many contemporary commentators and observers of the French Revolution, it seemed that the growing disenchantment with the absolutist regime of Louis XVI had been fostered in part by a particular kind of philosophy. French *philosophes* condemning the iniquities of the *ancien régime* drew parallels between the organization of society and the organization of nature. Like many other Enlightenment thinkers, they took it for granted that science, or natural philosophy, could be used as a tool to understand society as well as nature. They argued that the laws of nature showed how unjust and unnatural the government of France really was. It also seemed, to some at least, that the French Revolution provided an opportunity to galvanize science as well as society. The new French Republic was a tabula rasa on which the reformers could write what they liked. They could refound society on philosophical principles, making sure this time around that the organization of society really did mirror the organization of nature. Refounding the social and intellectual structures of science itself was to be part of this process. In many ways, therefore, the storming of the Bastille led to a revolution in science as well.

To many in this new generation of radical French natural philosophers, mathematics seemed to provide the key to understanding nature. This was nothing new in itself, of course. Greek philosophers such as Pythagoras had argued that nature could be comprehended mathematically. Far more recently, Galileo, Kepler, and Descartes, among others, had shown just what could be achieved by approaching nature through the language of mathematics. The hero of the mathematical worldview—in France as much as in his native England—was, however, Sir Isaac Newton. To French philosophers such as Condillac, Diderot, and Voltaire, Newton's *Principia* set the standard for the mathematical understanding of nature. Many late eighteenth-century natural philosophers—seeing themselves as following in Newton's footsteps—placed increasing emphasis on accurate measurement, on numbers, and on the development of powerful new mathematical analytical tools with which to manipulate their findings about nature. In revolutionary France, in particular, this new emphasis on the mathematical and the quantitative in natural philosophy was held up as a prerequisite for finishing Newton's task and producing a complete and final picture of an ordered and rational clockwork universe. Newton's French followers set aside Newton's vision of a universe in which God was continually present and refashioned his work as the epitome of Enlightenment rationalism.

Following the revolution, French scientific institutions were overturned, as ancient institutions as well as heads toppled to the ground. The royalist Académie Royale des Sciences was abolished and replaced with an Institut Nationale in 1795. At about the same time, new educational establishments such as the École Polytechnique and the École Normale were set up to train new cadres of revolutionary savants. Particularly following Napoleon Bonaparte's coup d'état of 1799, senior natural philosophers at the institute held positions of increasing political power. Napoleon was a keen advocate of, and enthusiast for, the physical sciences. The physicist Pierre-Simon Laplace, along with his close ally the chemist Claude-Louis Berthollet, held a firm grip on the reins of scientific power in France. Laplace, seeing himself as a committed Newtonian, wanted to complete what he regarded as Newton's grand project of reducing the universe to clockwork. French mathematicians such as Joseph-Louis Lagrange were producing powerful new mathematical techniques and applying them to understanding nature. Major strides forward across the board of rational mechanics were being made by the likes of Jean-Charles Borda, Réné-Just Hauy, and Laplace himself. Laplace promoted his allies and his protégés to positions of influence. His *Mécanique Celeste* was a manifesto of Newtonian science and an exemplar of how scientific progress should take place.

Radical young mathematicians in England such as Charles Babbage and John Herschel looked enviously on at the great strides achieved in revolutionary and Napoleonic France. Undergraduates at the University of Cambridge, they regarded their alma mater as a reactionary backwater—in both political and scientific terms. Cambridge's mathematicians were not only fervent Newtonians, but still wedded to Newton's way of doing mathematics as well. They had no time for the revolutionary gibberish being produced across the Channel. It was atheistic, materialistic, and (of course) French. Babbage, Herschel, and their cohorts vowed, however, to change all that and introduce analysis to Cambridge and to England. Cambridge professors and fellows devoted themselves to educating the sons of the gentry, preparing them to govern the burgeoning empire. They were taught mathematics because its rigors were held to be good training for the mind. Babbage and Herschel concurred with that at least. They wanted to make mathematical analysis into the foundation of a whole new understanding of the way the mind worked and how science could progress. They looked for links with business and commerce, moreover. The new science of analysis would lead to a proper understanding and organization of political economy as well.

By midcentury, Cambridge's mathematical reputation had been transformed. The university was probably the premier European institution in terms of mathematical training. Undergraduates underwent rigorous preparation in the latest mathematical techniques and their applications to physics before undergoing a grueling examination at the end of their student careers. Mathematics was still held to be a study calculated to breed gentlemen fit to govern an empire—a high ranking in Cambridge's mathematical league table could guarantee a successful career. Increasingly, however, ambitious young natural philosophers looked to the Cambridge mathematical Tripos (as it was called) as well, to provide them with a thorough grounding in the latest mathematics and its applications. Competition for the highest honor—the senior wranglership—was fierce. The university developed a unique culture of mathematics training designed to carry students through the rigors of the Tripos. Sporting prowess was encouraged as a means of relaxing the mind while turning the body into a fit receptacle. Two giants of nineteenth-century British physics—Maxwell and Kelvin—were products of the Cambridge system, as were a host of others. By the end of the century, Cambridge mathematical physics in many ways epitomized British science.

The German lands were developing their own culture of mathematical physics during the nineteenth century as well. The mid-nineteenth-century generation of German natural philosophers reacted strongly against what they perceived to be the metaphysical excesses of early nineteenth-century Romantic *Naturphilosophie*—as we shall see in more detail in subsequent chapters. Natural philosophy played an increasingly central role in German education as the century progressed. New research institutions were established with the express aim of placing German scientists at the forefront of natural philosophy. In many ways, theoretical physics as now recognized had its origins in these nineteenth-century German institutions. Physicists such as Rudolf Clausius in Zurich, Ludwig Boltzmann in Vienna, Bernhard Riemann in Göttingen, and Carl Neumann in Leipzig prided themselves on the abstractedness of their theoretical practice. Theory was a valid exercise in its own right. Where British natural philosophers worried about the material foundations of the terms and concepts deployed in their theories, German theoretical physicists by and large had no such concerns. What mattered was the integrity of the theory and its capacity to explain and predict the phenomena.

Physics in the nineteenth century was developing into strong and robust forms of practice, each with its own styles and traditions of research and institutional bases. As the century went on, success in physics increasingly came to be recognized as a marker of national status as well. Great men of science started to be recognized as national heroes as much as statesmen and soldiers. Laplace in France, Kelvin in Britain, and Helmholtz in Germany were national figures. Physics was becoming a way of fashioning oneself upon the national (and international) stage. In many ways the story of nineteenth-century physics is the story of the struggles of its practitioners to carve out a distinctive cultural niche for themselves and their way of doing things. For much of the nineteenth century there was no clear-cut, straightforwardly defined way of "doing" physics. There was no career pattern that the budding physicist might follow from school to university to research institution. Nineteenth-century physicists had to fashion themselves. They had to make up their careers as they went along.

## The French Revolution

French scientific institutions in the late eighteenth century were unmistakably part of the *ancien régime*. The country's premier scientific institution, the Académie Royale des Sciences in Paris, was the creature of royal patronage. There was nothing surprising therefore in the revolutionary

Committee of Public Safety's decision to suppress the Académie, along with other royalist institutions, including the universities. It was perceived as privileged, aristocratic, and elitist and opposed therefore to the ideals of the Revolution. When it was replaced a few years later in 1795 by the first class of the Institut Nationale, that new establishment was regarded as having a crucial role to play in furthering the Revolution and France's interests. Men of science were being mobilized for the war effort and came to play increasingly important roles in the Republic's affairs. This trend continued after Napoleon's takeover of the state. This forging of a new relationship between scientific institutions and the state provided the opportunity for some influential natural philosophers to implement their own particular visions of physical science. Pierre-Simon Laplace in particular took advantage of this chance to implement his grand Newtonian vision of a comprehensive physical theory that would lay bare the clockwork mechanism of the universe.

Physical astronomy, for Laplace, was the exemplar science. It was implicit in his view of science that all natural phenomena could be accounted for in just the same way that Newton had accounted for the movement of heavenly bodies. Just as the force of gravity dictated the movements of the stars and planets and of bodies on the Earth's surface, so could similar forces acting in the same way explain other kinds of movement. "By means of these assumptions," he asserted, "the phenomena of expansion, heat, and vibrational motion in gases are explained in terms of attractive and repulsive forces which act only over insensible distances . . . All terrestrial phenomena depend on forces of these kinds, just as celestial phenomena depend on universal gravitation. It seems to me that the study of these forces should now be the chief goal of mathematical philosophy."[1] Laplace's monumental *Traité de Mécanique Céleste*, in which these words appeared, was published in five volumes between 1799 and 1825. It contained a comprehensive manifesto of Laplace's vision of the end of natural philosophy in a unified Newtonian theory of everything. All of physical science could be reduced to the study of the force interactions between particles. The active powers of electricity, magnetism, heat, light, and so forth were to be understood as imponderable fluids made up of discrete particles interacting with each other in just the same way as the planets interacted with the Sun.

[1] P.-S. Laplace, *Traité de Mécanique Céleste* (Paris, 1799–1825), 5: 99, trans. in R. Fox, "The Rise and Fall of Laplacian Physics," 89.

Entrenched in a situation of ever increasing power and prestige within the Napoleonic French state, Laplace was in an ideal position to put his project into practice. He gathered a constellation of committed disciples around himself, all of them convinced like him that the holy grail of physics was to reduce everything to the interaction of particles in space. Laplace and his friend and fellow Bonapartist Claude-Louis Berthollet both owned country properties at Arcueil, a few miles south of Paris. There from 1801 onwards they organized the Société d'Arcueil, an informal society of their friends and protégés similarly committed to the project. They met there weekly to discuss their mutual interests in science and to plot their activities within the first class of the institute. The society provided Laplace and Berthollet with a secure base and a support structure from which they could engineer the elevation of their protégés into key positions within the powerful first class of the Institut Nationale and into influential teaching positions within the École Polytechnique and elsewhere.

French scientific institutions as reorganized under Napoleon were structured in a strict and centralized hierarchy. At the top of the pyramid were the prestigious members of the first class of the Institut Nationale. Membership in the institute was by election, and holders were salaried servants of the state. Members wielded a considerable power of patronage as well. Their say-so could be instrumental in determining the appointment of a budding young scientist to a salaried position teaching at one of the Parisian écoles or at a provincial university. Through his powers of patronage, Laplace was in a position to further the careers of his protégés; Jean Baptiste Biot and Étienne Louis Malus, students at the École Polytechnique, were promoted to positions of power and influence by Laplace. One function of the institute was the organization of prestigious prize competitions for significant new work in physics. Laplace was in a position to help ensure that prizes were awarded in areas of research in which his disciples were active. Thus, in 1807, for example, the first class of the institute proposed as a subject for the prize in mathematics a study of the phenomena of double refraction. Malus duly won the prize of 3,000 francs in 1808 with a theoretical extension of Laplace's own work on refraction in the *Mécanique Céleste*.

One of the exemplars of how to do Laplacian science was Laplace's own theory of capillary action. Trying to explain the tendency of a liquid in contact with a solid surface (such as the inside of a tube) to creep up that surface to some extent was a standard problem for eighteenth-century natural philosophers. It was axiomatic in Laplace's approach that such capillary action was the result of short range forces acting between the particles of liquid and the particles of solid. The issue was what form the law governing those forces should take. Some argued that it must be an inverse square law such as Newton had identified for gravitational force. Others argued that the inverse square law could be modified for inter-molecular distances. Laplace succeeded in sidestepping the dispute with an elegant mathematical demonstration showing that the precise form of the law was unimportant for its solution. Compared with previous efforts to solve the problem, Laplace's offering was lengthy and comprehensive. Along with his treatment of refraction, again based on the assumption that the phenomena were to be understood as the result of short-range in-teractions between particles—of light and solid matter in this case—this work not only supplied his allies and protégés with concrete examples of what a comprehensive theory should look like and how it ought to be constructed, but also provided them with a wealth of experimental work to confirm and expand Laplace's own hypotheses.

Étienne Malus's work during the 1800s on the reflection and refrac-tion of light beams fitted neatly into this picture. As a good Laplacian and faithful disciple it was axiomatic to Malus that light was made up of particles rather than waves. This was a moot point for eighteenth-century natural philosophers. The illustrious Newton was ambivalent on the matter, while the eminent Dutchman Christiaan Huygens had pro-duced solid results with a wave theory. According to the particle theory, or corpuscular theory, light consisted of a stream of particles emanating from the illuminated body and striking the eye of the observer. The theory was highly successful. By applying the laws of Newtonian mechanics to the light particles it was possible to explain a range of optical phenom-ena like reflection and refraction. Proponents of the wave theory, on the other hand, argued that light was the result of undulations in a universal, cosmos-filling immaterial medium. Vibrations in the illuminated body were transmitted like waves through this universal medium—or ether—to impinge on the observer's vision. Huygens applied the wave theory to provide rival explanations to those of the corpuscularians. He also suc-ceeded in explaining the curious phenomenon whereby objects viewed through crystals of Iceland spar appeared double—a phenomenon known as double refraction.

Malus succeeded, however, in reproducing Huygens's results and his explanation of double refraction using a corpuscular theory of light. He also made a major discovery. He found that light was polarized by reflection. In other words, light reflected from a surface appeared to be asymmetric—it acted differently along different directions. Polarization could be demonstrated by looking at light through particular kinds of crystals. If the crystal was held one way, the light source was visible. If the crystal was rotated by a right angle, the light source disappeared. This was a major triumph for the corpuscular theory of light. It seemed incompatible with the wave theory but easily explicable by assuming that the individual particles of light rotated around axes that were at an angle to their direction of motion. Malus's triumph was to reduce the various phenomena of reflection, refraction, and polarization to a single mathematical law, built around the assumption of asymmetric particles of light: "If we consider in the translation of the light molecules their motion around their three principal axes, a, b, c, the quantity of molecules whose b or c axes become perpendicular to the direction of the repulsive forces will always be proportional to the square of the sine of the angle these lines will have to describe about the a-axis in order to take up this new direction."[2] It was a classic piece of Laplacian science.

Laplace's success and that of his vision of a complete Newtonian philosophy of nature were closely tied to the fortunes of the Napoleonic state. While the empire flourished, so did Laplace. When the empire collapsed, however, so did the Laplacian empire of natural philosophy. Following Napoleon's fall at the Battle of Waterloo in 1815, Laplace lost a great deal of the political power he had wielded so effectively for the past fifteen years. Laplace's position under the restored Bourbon monarchy was by no means as secure as it had been before. His powers of patronage were increasingly curtailed along with his power to prevent political and scientific opponents from having their voices heard. In the years following the Restoration, therefore, an increasing number of Young Turks from a new generation set themselves up in explicit opposition to the Laplacian camp. Some of these rebels, such as Pierre Dulong and François Arago, were defectors from the Laplacian camp—both had been members of the Société d'Arcueil and had benefited from its patronage. Others, such as Joseph Fourier and Augustin Fresnel, were provincials who had had little previous contact with Laplace and Parisian scientific circles. Fourier, the oldest member of the burgeoning anti-Laplacian alliance, was a former army officer who had served under Napoleon in the disastrous Egyptian campaign. Fresnel was a known royalist sympathizer, a graduate of the

[2] E. L. Malus, "Théorie de la Double Refraction," *Mémoires Savants Étrangers*, 1811, **2**: 496, trans. in E. Frankel, "Corpuscular Optics and the Wave Theory of Light," 147.

École Polytechnique and the École des Ponts et Chausses who had spent most of his career in the provinces as a civil engineer.

Fourier had first come to Parisian scientific attention with a mammoth treatise on the distribution of heat in solid bodies, read out before the first class of the Institut Nationale in December 1807. Completely bypassing the standard Laplacian route of deriving his equations by treating the phenomena as the result of force interactions between particles of heat (the caloric theory of heat), Fourier developed his own way of approaching the problem, cultivating a whole new mathematical technology along the way. In his presentation, Fourier was determinedly agnostic concerning the physical nature of heat, preferring to focus on a more abstractly mathematical formulation of the problem. A few years later he successfully submitted a revised version of his treatise for one of the first class's prize competitions, judged by a panel including Laplace himself as well as Lagrange, Legendre, Malus, and Hauy—all good Laplacians. Despite that achievement, publication of his work was blocked and his prize-winning contribution did not appear in full until 1823, long after Laplace's fall from grace and when Fourier himself was already permanent secretary of a revived Académie Royale des Sciences. In the meantime, an extended abstract of his work was published by the sympathetic anti-Laplacian, Arago, in the *Annales de Chimie et Physique* of 1816.

More trouble came to the Laplacians from the field of optical theory, so recently hailed as the site of some of their greatest triumphs in the form of Laplace's work on refraction and Malus's discovery of polarization. This work had been extended by both Jean Baptiste Biot and François Arago in classic Laplacian fashion. Before long, however, the two experimenters fell out in a public and acrimonious priority dispute in which Arago accused Biot of appropriating his discoveries for himself. Disenchanted with the Laplacians, Arago was more than happy to place his considerable influence at the disposal of Augustin Fresnel when the young outsider tried to interest Parisian savants in his own rival wave theory of the phenomena of diffraction—the breaking up of a beam of light into a series of light and dark bands when it passes through a narrow slit or past the edge of a body. Fresnel explained diffraction by supposing that dark bands were caused by the coincidence of peaks and troughs in waves of light canceling each other out at particular points along the wave front while light bands were the result of two peaks reinforcing each other. Having arrived in Paris in the summer of 1815, Fresnel was soon in contact with Arago, who took him under his wing and pointed him in the direction of the Englishman Thomas Young's studies on the wave theory of light. Arago undertook, moreover, to act as reporter for Fresnel's first presentation before the first class of the institute later that year. As well as submitting a highly complimentary report, Arago also had Fresnel's memoir published in his *Annales de Chimie et Physique* and succeeded in finding him a permanent position in Paris so that he could continue his researches. The turncoat Arago was turning the Laplacian patronage network against itself.

The Laplacians threw down the gauntlet. In 1817 a commission of the academy, packed with Laplacians, called for a prize competition on the subject of diffraction. Their hope was to repeat the triumph of 1808, when Malus had carried off the laurels with his corpuscularian study of refraction. There was even a staunch Laplacian, Claude Pouillet, one of Biot's students, working on the problem. The terms in which the competition was posed made it clear that the commission expected a corpuscularian victor. The commission, however, underestimated Fresnel, who responded with a revised treatise, ironing out problems in his original presentation and producing new mathematical laws for deriving the phenomena of diffraction. The commission, despite its corpuscularian bent, had little choice but to award him the accolade. A few years later, even Biot, a committed proponent of the corpuscularian theory, was admitting that "the principle of interference is, up to now, the only one with which one can explain the particularities of diffraction, and in that this phenomenon is favourable to the undulatory system." He still maintained though that there was something unsatisfactory about the solution: "one feels that it offers rather a representation of the phenomena than a rigorous mechanical theory."[3] In due course he hoped that it would be replaced by a suitably materialist, that is to say corpuscularian, theory.

Biot was to be disappointed. One by one the citadels of Laplacian physics fell before the interlopers. By the 1820s, both the caloric theory of heat and the Laplacian two-fluid theories of electricity and magnetism were under fierce attack as well. In mechanics, orthodox Laplacian "physical mechanics" was being replaced by the "analytical mechanics" practiced by Fourier and his protégés such as Claude Navier and Sophie Germain. Laplace's ally Simeon-Dénis Poisson deplored the new style and hankered for the days when mathematicians would "re-examine the leading problems of mechanics from this point of view, which is at once

[3] J. B. Biot, *Précis de Physique*, 3rd ed. (Paris, 1824), 2: 472–73, trans. in E. Frankel, "Corpuscular Optics and the Wave Theory of Light," 162.

physical and consonant with nature."[4] Abstract analysis was all very well, but what was really needed was a mathematics that stayed in touch with material reality. As the Laplacian generation either grew older or fell out of political favor under the restored monarchy, their positions in the seats of power were usurped by their political and scientific opponents. Arago, Ampère (another anti-Laplacian), and Fourier were already members of the reconstituted Académie Royale des Sciences. Fresnel was elected in 1823. A year earlier, in 1822, Fourier had scored a decisive victory by trouncing Biot in the election for one of the permanent secretaryships. The loss of political power and the loss of scientific credibility appeared to go hand in hand.

Laplace had presided over a remarkably productive two decades of science in France. His brand of revolutionary science had brought about a transformation in the eighteenth-century Newtonian synthesis. As far as his adherents were concerned, his magisterial *Mécanique Céleste* provided the blueprint for a thoroughgoing Newtonian overhaul of physics. It showed as well how successful mathematics could be as a tool with which to comprehensively interrogate nature. This new, sweeping, and powerful science went hand in hand with the revolutionary reform of France. Its uncompromising materialism fitted in well with the guiding philosophy of the newly dominant elite. The Revolution and its Bonapartist aftermath gave Laplace the political clout to put his vision into practice as well. Laplace had the power to hire and fire. He could put into positions of influence those who shared his commitment to a materialist reading of Newtonianism. The shake-up of French scientific institutions after the Revolution and under Napoleon's dispensation allowed for scientific careers for the talented in a way that few had previously been able to aspire to. To envious eyes beyond the boundaries of the Empire, French science could easily appear as an ideal to be fondly emulated.

### The Analytics

English science at the turn of the century—at least in its upper echelons—was very much an aristocratic affair. A smattering of natural philosophy was part of the cultural repertoire of leisured gentility. Practicing natural philosophers were often either gentlemen themselves, with the time and

---

[4]S. D. Poisson, "Mémoire sur l'Équilibre et le Mouvement des Corps Élastique," *Mémoires de l'Académie des Sciences*, 1829, 8: 361, trans. in R. Fox, "The Rise and Fall of Laplacian Physics," 118.

resources to devote themselves to science, or men beholden to such gentlemen for patronage. English men of science and fellows of the Royal Society prided themselves on their scientific heritage. They were after all the inheritors of the great Sir Isaac Newton's mantle. The president of the Royal Society in 1800—Sir Joseph Banks, who had made his name as a botanist on Captain Cook's voyages of exploration in the South Seas—had already been at the helm for more than twenty years and was to remain there for another twenty. As his reign lengthened, more and more of the younger generation of natural philosophers became restive—particularly those interested in the mathematical sciences of which Banks (reputedly at least) disapproved. They regarded English science as becoming ever more backward and reactionary, losing touch with the developments taking place in Continental Europe. They abhorred Banks and his patronage networks and wanted science to be a meritocracy instead. They wanted to forge links between science and commerce rather than kowtow to lords and ladies of leisure.

Early nineteenth-century Cambridge remained a bastion of academic and aristocratic privilege. Its students were largely drawn from the ranks of the landed gentry and the university's purpose was to provide the finishing touches to their education, to prepare them for service to church or state. Mathematics was perceived as having a central role to play in achieving this end. It provided an unparalleled means of training the mind. A student who could follow the complexities of Euclid's geometry or Newton's fluxions (as Newton's style of calculus was called) was judged to be capable of following a course of reasoning in other walks of life as well, be it the law, politics, or theology. Cambridge's mathematical professors and scholars had little time for new developments. The university was a citadel of learning, not of research. They preferred to follow tried and tested methods rather than dabbling with dubious (and foreign) innovations. As befitted its status as nurturer of the nation's future elite, the university, particularly during the Revolutionary and Napoleonic wars, was politically and theologically conservative as well. Students had to swear their allegiance to the thirty-nine articles of the Church of England before graduation; heresy both in politics and in theology was firmly stamped down.

For a new breed of student in early nineteenth-century Cambridge, however, this state of affairs was deeply unsatisfactory. Men such as Charles Babbage and John Herschel admired French science and politics and were deeply contemptuous of what they regarded as the culpable ignorance and reactionariness of the Cambridge dons. Babbage was the son of a wealthy banker. Herschel, of course, was the son of the celebrated Hanoverian emigré, the musician and astronomer William Herschel, discoverer of Uranus. Both had republican sympathies. Babbage was acquainted with Napoleon's exiled younger brother, Lucien. Herschel had visited Paris with his father and been introduced to Napoleon himself. In letters to Babbage he addressed him as "citizen" in the French fashion and after Waterloo expressed disquiet to his friends as to his future as a "poor snivelling democratic dog"[5] in a world dominated by triumphalist aristocrats. Along with others such as George Peacock, Alexander d'Arblay, and Edward Ffrench Bromhead, they mixed their enthusiasm for revolutionary politics with a taste for revolutionary mathematics. Disdaining the Newtonian bias so prevalent in Cambridge, they immersed themselves instead in the latest products of French analysis. In their politics and their intellectual allegiances they stood for everything that stalwarts of the Cambridge regime such as Isaac Milner, the redoubtable president of Queens' College, found abhorrent.

The outcome of their backroom conspiracies was the foundation of the Analytical Society in 1811, committed to introduce French mathematics into the University of Cambridge. The society was started almost as a joke—a joke at the expense of the university's conservative politico-theological wranglings. A dispute was raging over the foundation of a Bible Society. While some argued that the Bible should be circulated along with the Book of Common Prayer to guard against heretical misinterpretations of the word of God, others were adamant that the Bible should be distributed alone. It was a dispute as to the extent the poor could be trusted to read God's word unsupervised. In the midst of this furor, Babbage in his rooms at Trinity College drew up plans for an alternative society. It was to be established to support the publication of the French mathematician Silvestre François Lacroix's *Differential and Integral Calculus* in English, a work, according to Babbage's broadside, already "so perfect that any commentary was unnecessary."[6] The lampoon did have a serious intent, however. Babbage and his cohorts—all high-flying mathematicians aiming at high honors in the mathematical Tripos—were disgusted by the state of affairs at Cambridge. They wanted to revolutionize the Tripos and bring it, as they saw it, up to date.

---

[5]J. Herschel to J. Whittaker, 7 July 1815, quoted in H. Becher, "Radicals, Whigs, and Conservatives," 411.

[6]C. Babbage, *Passages from the Life of a Philosopher* (1884; reprint, London: Pickering & Chatto, 1991), 20.

The society's aim was to support "the Principles of pure D'ism in opposition to the Dot-age of the University."[7] The slogan was a barbed in-joke and a pun. The new French analytical calculus employed the now conventional notation $dx/dy$. The university's favored approach was that of Newtonian fluxions, which would express the same concept as $\dot{x}$. At the same time, the radical theological principles of deism (denying Revelation and the Trinity) were being opposed to the university's muddleheaded dotage. Babbage and Herschel were the new society's leading lights, both committed to the new system. They consumed French mathematics voraciously and produced their own contributions prodigiously, published in their own in-house journal, the *Memoirs of the Analytical Society*. Both, along with Peacock, had eyes on a Cambridge fellowship. Babbage flunked, however. Having moved to Peterhouse from Trinity so that he might have a chance of a fellowship without taking holy orders, he fell afoul of the university's religious ordinances and thus was unable to aim for honors. He got an ordinary degree without examination and lost his chance of a college career. Herschel graduated senior wrangler at St. Johns in 1813, gaining a college fellowship. Peacock came second and gained a fellowship at Trinity, accepting ordination along the way.

Within a few years, George Peacock was appointed one of the university's examiners and took advantage of his position to start introducing the new, infidel "d-istic" notation into the Cambridge examination papers. The analytics' logic was simple—if the new mathematics were in the examination papers, then Cambridge's private tutors, who undertook the bulk of teaching, pragmatists to a man, would start teaching it to their students. Its introduction in 1817 was highly controversial to say the least. Looking back, Peacock suggested that only the success of students from St. John's at the examination (the master of St. John's was vice-chancellor of the university that year) prevented him from being hauled in front of the university courts for his temerity. The opposition had some real intellectual concerns about the new mathematics. As George Peacock's opponent Daniel Peacock (no relation) put it, "Academical education should be strictly confined to subjects of real utility, and so far as the lucubrations of the French analysts have no immediate bearing on philosophy, they are as unfit subjects of academical examination, as the Aristotelian jargon of the old schools."[8] The complaint was that French

---

[7] Ibid., 21.

[8] D. Peacock, *A Comparative View of the Principles of the Fluxional and Differential Calculus* (Cambridge, 1819), 85.

analysis lost its grip on reality. Powerful it might be, but its symbols did not refer to anything in the real world. Its techniques simply provided a shortcut through a problem without providing the kind of intuitive, if plodding, understanding that an undergraduate needed if he were to have his mind trained for empire.

For Herschel and Babbage, however, there was more to analysis than a debate about the appropriate mathematical symbols, or the proper education of Cambridge undergraduates, or even mathematics itself. Analysis was part and parcel of a grand project of intellectual, economic, and cultural reform that they hoped would turn British society on its head. They agreed with the Cambridge dons that mathematics was preeminently a way of training and organizing the mind. They differed, however, in their methods and in what they wanted the mind trained for. These were representatives of the new urban industrial middle class. They saw Britain's future in industrial expansion and the thoroughgoing application of political economy. The key to the success of analytical algebra as they saw it was its efficiency. It was a problem-solving technology that could produce answers quickly and without wasting resources. That was why it was good mental training. It exemplified efficiency. More than that—it mirrored the workings of an ideal mind as well. It was a way of economizing mental labor. As such it could be used to recognize what the most efficient way of proceeding in other enterprises might be too. It could provide the key, for example, to the most profitable way of deploying resources in order to maximize factory production.

Following his enforced departure from Cambridge, Babbage switched his field of operations to the metropolis. There, his enthusiasm for finding ways of maximizing the efficiency of mental labor in the same way that the division of labor was increasingly being deployed to maximize the efficiency of manual labor earned him a receptive audience. London's bankers and industrialists were as keen as he was to find ways to improve their balance sheets. Babbage's circle in London included men such as the stockbroker Francis Baily and the actuary Benjamin Gompertz. Both were enthusiastic mathematicians and astronomers, convinced, like Babbage, that their science could and should be prosecuted like their business—and vice versa. Efficiency was the name of the game in both cases, and efficiency was best achieved by due attention to, and proper application of, the laws of nature and the operations of the mind. Babbage, Baily, and Gompertz, as well as Herschel, were instrumental in establishing the Astronomical Society in 1820 as an alternative power center to Sir Joseph Banks's corrupt (as they saw it) domination of the Royal Society.

Following Banks's death in office in 1820 and throughout Sir Humphry Davy's precarious presidency of the Royal Society during the 1820s, radicals, spearheaded by Babbage and his Astronomical Society cohorts, battled with the conservatives for control of the Royal Society and its near monopoly of governmental patronage for science. The battle culminated in John Herschel's unsuccessful stand against the duke of Sussex (the king's younger brother) for the Royal Society's presidency in 1830.

This was a battle about efficiency and the proper division of labor in science. The problem with the Banksian regime and its successor, in Babbage's and his friends' minds at least, was that it interfered with the proper and transparent workings of the scientific community. It depended on backroom backhanders instead of meritocracy. The superiority of algebraic analysis over geometrical reasoning lay in its efficiency and transparency as well. Babbage argued that "[t]he power which we possess by the aid of symbols of compressing into a small compass the several steps of a chain of reasoning, whilst it contributes greatly to abridge the time which our enquiries would otherwise occupy, in difficult cases influences the accuracy of our conclusions: for from the distance which is sometimes interposed [in geometrical reasoning] between the beginning and the end of a chain of reasoning, although the separate parts are sufficiently clear, the whole is often obscure. This observation furnishes another ground for the preference of algebraic over geometrical reasoning."[9] Not only was analysis more efficient, it was less prone to error than geometry—it was easier to scrutinize. That kind of oversight, according to Babbage, was the key to good science and the key to good management in both industry and science.

Babbage's ultimate solution to the problem of how to guarantee efficiency, transparency, and accuracy in reasoning was the same as his solution to the same problem in political economy: replace humans with machinery. Babbage was a firm exponent of the division of labor in factory management and equally enthusiastic for mechanization as the ultimate realization of the principle. His primary concern throughout the 1820s and beyond was to work on his projected calculating and analytical engines and to persuade a sometimes reluctant government to finance the project. The calculating engine would replace the human drudge work of calculating mathematical tables to be used (for example) in actuarial work and in astronomy. The analytical engine would go further—it

---

[9]C. Babbage, "On the Influence of Signs in Mathematical Reasoning," M. Campbell-Kelly (ed.), *The Works of Charles Babbage* (London: Pickering & Chatto, 1989), 1: 376.

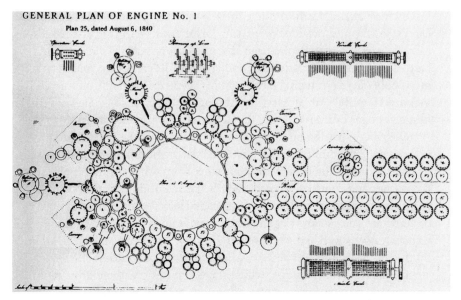

GENERAL PLAN OF ENGINE No. 1

Plan 25, dated August 6, 1840

2.1 Plans for Charles Babbage's ambitious Analytical Engine, showing details of its inner
mechanism. Babbage argued that by finding a way of mechanically reproducing the
mental attributes of memory and foresight he could build an intelligent machine that
could be used to replace monotonous mathematical labor.

would replace the human capacity to reason as well (figure 2.1). "Memory and foresight," according to Babbage, were the foundations of human
intelligence, and he had found a way of embodying them in a machine.
Memory was achieved by the "principle of successive carriages." Foresight was more difficult. Babbage recalled triumphantly that "[i]t cost me
much thought, but the principle was arrived at in a short time. As soon as
that was attained, the next step was to teach the mechanism which could
foresee to act upon that foresight. This was not so difficult: certain mechanical means were soon devised which, although very far from simple,
were yet sufficient to demonstrate the possibility of constructing such
machinery."[10] His analytical engine was to be the final realization of the
analytics' dream of industrializing the operations of the human mind and
the scientific community along the same lines as the industrialization of
the economy.

[10] C. Babbage, *Passages from the Life of a Philosopher* (1884; reprint, London: Pickering &
Chatto, 1991), 46.

Herschel's failure to defeat the duke of Sussex in the 1830 election for the presidency of the Royal Society, along with his own continuing difficulties in acquiring government financial support for his calculating engines, lay behind Babbage's publication in 1830 of his controversial *Reflections on the Decline of Science in England*. The book was a passionate broadside against the corruption, mismanagement, and nepotism of English science in general and of the Royal Society in particular. The Royal Society needed a complete overhaul so that it could be recognized as the legitimate overseer of the division of scientific labor between the growing number of specialist scientific societies (like the Astronomical Society) and the proper allocation of government resources. Babbage's model for future reform was unambiguous. He had his eye on the power and prestige of the Académie des Sciences across the Channel. Its officers were salaried servants of the state and had the financial and political clout to push science forward. Babbage saw this centralized, Bonapartist monolith as an antidote to corruption and the epitome of efficient management. Others, even among his fellow reformers, disagreed of course, pointing out that the academy was even more prone than the Royal Society to corruption and backroom power broking. It seemed obvious to Babbage, however, that the importation of French science and French scientific structures should go hand in hand.

Babbage's, Herschel's, and the rest of the analytics' apparently local battle to introduce French analysis into Cambridge's moribund mathematical culture and the fierce opposition their efforts encountered were symptomatic of broader battles within the world of English science. Young Turks such as Babbage and his cronies wanted to turn English science upside down and remake it in their own image. New methods of mathematical analysis, bizarre as it may seem to modern readers, were a way of trying to achieve this. These men saw analysis as encapsulating a new and more efficient way of thinking that could be applied outside the narrow confines of the university and its hidebound curriculum, just as it could be used to drag that curriculum and its guardians screaming and kicking into the new century. Cambridge was not a bad place to start the battle, since its professors presided over the education of a large portion of the country's future ruling elite. Efficiency and meritocracy were the buzzwords of an increasingly confident new industrial class that was just embarking on its own campaign for political power to match its growing economic clout. The analytics and their analysis were in the vanguard of that campaign.

## Cambridge Culture

By the end of the nineteenth century, the University of Cambridge was internationally recognized as a powerhouse of mathematical physics. Its former students could be found staffing new universities in Britain, throughout the Empire, and beyond. The place had become a veritable factory production line of mathematical physicists. This clearly was a huge change from the state of affairs that so depressed the Analytical Society in the 1810s. Indeed, their drive to reform the Cambridge Tripos was partially responsible for the transformation of the university's international reputation. There was more to it than that, however. The examination regime and the regime of mathematical training developed in Cambridge during the first half of the nineteenth century were quite explicitly designed to churn out mathematically adept individuals in large numbers (figure 2.2). The aim was not to produce mathematicians or mathematical physicists as such. Mathematics was taken to be a means of inculcating a rigorous education of the mind just as it had been earlier in the century. Cambridge products were meant to be fit to govern an expanding Empire that required their services in ever increasing numbers.

2.2 A Cambridge examination taking place in the Great Hall of Trinity College around 1840. Regimented and closely invigilated examinations like these are common nowadays but were a relative innovation at the time.

A candidate who was successful in the Tripos was taken to have demonstrated precisely those virtues of self-discipline, mental rigor, and iron determination that were assumed necessary to be capable of such service.

By midcentury, the Cambridge system of examination that Babbage, Herschel, and friends had considered so inadequate had undergone a major overhaul. In fact, the process had been under way for some time when they were undergraduates. From the late eighteenth century onwards, the emphasis in assessing students' ability gradually shifted from oral to written examination. Mathematics—the only subject to be examined formally and through which a student could attain honors—became increasingly important as a topic of study. Honors students were divided into three classes: wranglers, senior optime, and junior optime. Within these divisions, the examiners developed ever finer means of discrimination aimed at individually ranking each candidate for honors in the Tripos. Graduating as senior wrangler (first in the list of wranglers) or indeed as second or third wrangler was considered a major achievement. The analytics' efforts to introduce new mathematical styles and techniques into the syllabus had a major effect on the system. Increasingly, examiners developed finer means of grading questions so as to discriminate between different levels of ability. William Whewell, the polymathic master of Trinity College, played a major role during the 1830s and 1840s in reforming and rationalizing the Tripos system. As the examination system became ever more rigorous and taxing, submitting oneself as a candidate for honors meant being prepared to undergo a grueling and arduous regime of training.

The key to success in this punishing process was the acquisition of a well-established and successful personal tutor—or "coach," as they were popularly known. As the examination process became more demanding and punishing, the role of the university's own professors in the pedagogical process became less significant. After all, every student had easy access to their lectures. What was needed to gain an edge was a personal tutor with a proven track record of producing high wranglers. Coaches worked with their own chosen "teams" of students, inculcating tried and tested ways of approaching problems speedily and reliably. The teams worked their way through example after example of problems, internalizing the best ways of getting through the examination successfully, answering as many questions correctly in as short a time as possible. The best students aiming at the top few places in the lists needed to demonstrate considerable flair, ingenuity, and originality to attain the high honors they hoped for. This could be achieved only if they had the mathematical techniques required to solve the examination questions at their fingertips. Coaches aim was to drum such techniques into their disciples' heads by constant repetition and exercise. A good candidate was expected to be able to read a question, recognize the techniques required for its solution, and apply them successfully while barely thinking about the matter.

Unsurprisingly perhaps, such an arduous and in many ways unprecedented regime of mental training had its failures. The road to Tripos stardom was littered with casualties. Even candidates who excelled at the Tripos recorded their dismay at the mental and physical strain they had been subjected to. Leslie Ellis, senior wrangler in 1840, recorded in his journal his "bitter dislike of Cambridge and my own repugnance to the wrangler making process."[14] Cambridge had developed its own solution to the problem of mental breakdown during the course of Tripos preparation, however. As they exercised their minds, Tripos candidates were encouraged to exercise their bodies as well. From the 1810s onwards, as the analytical revolution gathered pace and grinding application became more and more a prerequisite of Tripos success, hard physical exercise as an adjunct and antidote to rigorous study became commonplace. Cambridge was developing a culture of "work hard, play hard"; solitary activities were discouraged to prevent undue introspection. Students entered into sporting activity with as much self-discipline and rigor as they applied to their mathematical studies. Sport at Cambridge was not just the preserve of the idle aristocrats who had no interest in submitting themselves to the rigors of the Tripos examination. It was part and parcel of the university's mathematical culture.

Much of the university's culture by the late nineteenth century was built around the mathematics Tripos. The most successful candidates, who filled the highest positions in the rankings, were lionized not only within Cambridge but nationally as well. Their images and their histories would appear in the popular press. Their future careers would be assured. The awarding of degrees was hedged in by ritual. The results of the Tripos examination each year were publicly read out at the university's Senate House in strict order of ranking. Colleges vied with each other for the honor of the highest number of wranglerships. The senior wrangler each year would be carried from the Senate House on the shoulders of his peers and paraded around the city streets. Failure was accorded its ritual as well. The candidate achieving the lowest result each year was awarded the wooden spoon. The "spoon," fashioned, ironically enough, from a boating oar, would be lowered from the Senate House's galleries down to

---

[14]Quoted in A. Warwick, "Exercising the Student Body," 298.

the unfortunate recipient below. Success in the Tripos was a guarantee of entry into the country's cultural elite. Comparatively few high wranglers became professional mathematicians or men of science—that, after all, was not really what the Tripos was about. Those that did however, were sure of a head start.

Relatively few eminent British men of physical science during the second half of the nineteenth century had not passed through the Cambridge Tripos. William Thomson and James Clerk Maxwell are only the most eminent examples. They were joined by Peter Guthrie Tait, George Gabriel Stokes, George Bidell Airy, Lord Rayleigh, Joseph Larmor, and J. J. Thomson among others. These men and others like them contributed to constructing a distinctive style of mathematical physics in the second half of the century. It was a style that owed a great deal to their original training in the mathematics Tripos. Despite the analytics' revolution during the 1810s and 1820s and the consequent introduction of French and Continental methods of analysis, the Cambridge system still maintained a strong commitment to the traditional concern with "mixed mathematics." Examiners (and coaches) encouraged students to work on mathematical problems with a strong physical component. Mathematics was expected to describe and solve problems in the real world. Challenging questions in the Tripos examination often formed the basis for ambitious students' future research. The late nineteenth-century articulation of mathematical theories of the electromagnetic ether, for example, was very much a product of this Cantabrigian approach. Even much of the early twentieth-century British response to Einstein's newfangled theories of relativity was firmly grounded in this tradition of mathematical research.

Cambridge's mathematical culture during this halcyon period was, like the university's culture more generally, avowedly masculine. Mathematics was unambiguously men's business. As women were grudgingly admitted into the university's lecture theaters during the second half of the century, they were even more grudgingly admitted into the coaches' teams without participation in which they had no hope of achieving honors. Even when women were allowed to participate in the Tripos from the 1870s onwards, they were excluded from the public ranking system for several years. It caused a major scandal in 1890 when Phillipa Fawcett from Newnham College actually beat that year's senior wrangler. Not only did women's success bruise male egos and undermine the cultural kudos attached to mathematical preeminence, it also severely challenged views of the relationship between mathematicians' bodies and their minds. Athleticism mattered to Cambridge wranglers because it was held that a balance was needed between energies devoted to mental and those devoted to physical exertion. Such a balance was impossible in women's bodies since their physical energies were meant to be overwhelmingly directed towards maintaining their reproductive organs. They were thus judged incapable in principle of the rigorous mental work required for Tripos success. Cambridge mathematical physics itself, in the form of the doctrine of the conservation of energy, underpinned this model of bodily economic management.

By the end of the century, the Cambridge mathematics Tripos was under attack from reformers once again. Women were showing themselves quite capable of playing the game; this did nothing to help those who defended the Tripos as the preeminent means of sorting out the men from the boys. New centers of excellence in research and training were emerging as well, challenging Cambridge's claim to provide the best. Even within the university, the mathematics Tripos's position as the route to success in the physical sciences was being challenged by the natural sciences Tripos and the increasingly important role of the Cavendish Laboratory as a center of research. The popularity of German models of theoretical physics could be seen as a potent threat and a challenge to the hegemony of Cantabrigian mathematical physics in the "mixed mathematics" tradition. For much of the century however, Cambridge was acknowledged as one of Europe's most prolific producers of physical scientists. Shared experiences as fodder for Cambridge's wrangler mills and common ground in shared techniques and practices produced a highly cohesive and productive scientific elite that dominated physics for a large part of the second half of the century.

## The Reign of Theory

German natural philosophy and its institutions underwent their own reformation during the nineteenth century. The German lands at the beginning of the century were a patchwork of states, each with its own local university. By the end of the century, a unified Germany was one of the most powerful countries in Europe, its economy threatening to overtake that of Great Britain. Germany had universities to match its political and economic clout as well. In the sciences particularly, German universities increasingly looked world-beating. The new state placed great emphasis on scientific and technical education for its citizens, not only seeing science and technology as the foundations of its burgeoning economic power, but seeing scientific prestige as reflecting glory on the country that had produced it. Internationally recognized German men of science such as Hermann von Helmholtz and Emil du Bois Reymond were people to be reckoned with on the cultural and political scene as well. Their views mattered. Just as radical young natural philosophers in England at the beginning of the century cast envious eyes over the Channel at French scientific institutions, those calling for a new dispensation for British science and its institutions at the end of the century pointed to Germany as their model. Germany was a country that recognized the increasingly important role of science in its struggle for economic supremacy. British failure to emulate its institutions would be a recipe for British industrial decline.

As in England (and in France for that matter), universities in the early nineteenth-century German states had as their aim the education of the country's professional and ruling elite. German states usually had at least one university for this purpose—those that could afford them had more. As institutions they were designed to provide their privileged students with the kind of education that would mold them into future leaders of society. They would produce lawyers, medical men, teachers, and clerics. Typically, each university was divided into four faculties—law, medicine, theology, and philosophy. The first three of these provided particular professional training. Philosophy included the humanities, mathematics, and natural philosophy. Philosophy in particular was seen as providing for students at university level the opportunity to develop *Bildung* that was a major purpose of education. *Bildung* was in many ways the German equivalent of the English ideal of a liberal education—a training of the mind that was meant to produce depth and discrimination. At many German universities, this was achieved through immersion in a classical education (as it was at Oxford). Reformers argued, however, that another route to *Bildung* existed that did not involve the recapitulation of ancient knowledge. It could be achieved through science and mathematics.

Natural philosophy's status within philosophy faculties at the beginning of the century—and hence within German universities as a whole—was low. It was not considered as essential training for a profession such as medicine, theology, or the law. Neither was it regarded as an important component in the development of *Bildung*. Natural philosophy teaching took place as part of a general education curriculum that was not dissimilar to what students at schools and *Gymnasien* might experience. In no sense was a capacity for research considered a prerequisite for a university professorship in natural philosophy. The function of universities in general was pedagogy rather than the production of new knowledge.

When Carl Friedrich Gauss was asked by the University of Göttingen about his opinions concerning the filling of their vacant professorship of physics in 1831, he emphasized that the successful candidate's main obligation would be to deliver accessible lectures to a mixed audience who would only want to acquire a general knowledge of the subject. The candidate would also be a member of the Göttingen Society of Sciences and would therefore be expected to be proficient in mathematics and capable of producing work that could be published in the society's *Transactions*. Even for an eminent man of science such as Gauss, however, this was a secondary consideration. If anything, Gauss argued that a first-class mathematician would be unsuitable as he would be incapable of appealing to a broader audience through his lectures.

By the later 1830s and 1840s, however, the pedagogical status of natural philosophy teaching was changing. As early as 1824, for example, the curator of the University of Heidelberg proposed to the state of Baden's interior ministry that a "mathematical seminarium" should be established at the university. It was to be modeled on the increasingly popular and successful seminars in philology that were being credited with improving German classical education. Other universities followed suit. By the late 1820s, the University of Halle had a "physical seminar" in place, and when Wilhelm Weber, a student at Halle, was appointed to the professorship at Göttingen, he brought the model with him. The seminar model provided more intensive training, primarily aimed at improving the quality of schoolteachers. As Moritz Stern, extraordinary professor of mathematics at Göttingen argued in 1849, "The philological seminars came into being in an intellectual epoch . . . in which the knowledge of antiquity was seen as the almost exclusive foundation of all scientific knowledge. But the louder the so-called realistic direction demands its right, the greater the need becomes for all educated people to understand the foundations on which rest the mechanical and physical discoveries and inventions that affect our conditions so mightily, and the more it also becomes necessary that future teachers of mathematics and physics be offered an academic institute that has their further training as its special purpose."[15] Gradually, it came to be understood that some grounding in physical research might form a part of that "further training." The foundation of the Berlin Physical Society in 1845 was an augury of things to come.

Research increasingly came to be regarded by German natural philosophers as a potential route to prestige and career. The *Annalen der Physik*

---

[15]Quoted in C. Jungnickel and R. McCormmach, *Intellectual Mastery of Nature*, 1: 79.

devoted more and more of its pages to the productions of native men of science rather than to translations of works from foreign journals. Founded in 1790, the *Annalen* was by the 1840s the premier German journal of physical science. Its editor since 1824, Johann Christian Poggendorff, had the power to make or break a scientific career. Increasingly, research publication, and publication in the *Annalen* in particular, came to be regarded as a prerequisite for any hopeful candidate for a university professorship. Research was coming to be regarded as a value in its own right. Such a cultural sea change had a clear impact on the status of physics research in Germany. Researchers were no longer enthusiastic individuals or wealthy dilettantes with the leisure to indulge in experimental or mathematical tinkering. They were hardheaded professionals with all the prestige of state-salaried university professors. Research was turning into a career and increasingly prestigious German physics professors could demand ever more from their academies in terms of resources and facilities as rival institutions battled for their services.

Carl Friedrich Gauss was one of the towering figures in this transformation of German physics. As professor of higher mathematics and of astronomy at the University of Göttingen he had been instrumental in securing an institutional niche for research there during the 1830s. His increasing reputation as a mathematician and astronomer—particularly his international collaborations in astronomy and geomagnetic observations—raised the profile of research throughout the German lands. Gauss also established his own style of mathematical investigation in physics, particularly electromagnetism, that served as a crucial resource for the next generation. Bernhard Riemann had studied mathematics under Gauss during the 1840s, attracting his patronage along with that of Weber. Gauss encouraged his mathematical researches and his efforts to establish mathematical connections between previously disparate areas of physical inquiry. Riemann's work on electrodynamics in turn inspired another young German mathematician, Carl Neumann. Neumann's work in electrodynamics, along with that of Gauss, Riemann, and Weber, had a profound impact in establishing a crucial German theoretical presence in discussions of electromagnetism during the second half of the century. Electromagnetism—the science of the telegraph cables so crucial to imperial expansion—was, as we shall see, in many ways at the core of nineteenth-century physics.

Poggendorff continued editing the *Annalen* until his death in 1877. By this time his name had become synonymous with the flagship journal of German physics. Following his departure from the scene, the Berlin

Physical Society took the journal under its auspices, with Gustav Wiedemann as editor, advised by Hermann von Helmholtz on theoretical matters. A clear division of labor was emerging between the experimenter and the theoretician. The discipline of theoretical physics—a distinctively German institution in its origins—was taking off. German theoretical contributions to the *Annalen* were increasingly autonomous, contributors referred more and more to the theoretical contributions of fellow Germans as opposed to work from France or Britain. This is not to suggest that such work was ignored. Work by Faraday and Maxwell in particular was heavily drawn upon. It does show, however, the development of a distinctively German culture of theoretical physics with its own concerns and direction. The institutional structure of German science increasingly encouraged research. Directors of new physics institutes in particular had the time and resources to devote themselves to new theoretical investigations. Rather than being an adjunct to teaching, research by the 1860s or 1870s was regarded as being an end in itself. Institutes, their directors, and their students were state supported since their research contributed to the cultural prestige of the state itself.

From the 1870s in particular, extraordinary professorships of theoretical physics were established at a number of German universities. They were typically set up as junior positions in conjunction with the already established ordinary professorship of physics. The aim as a rule was to provide an additional source of teaching that would free the holder of the senior appointment to carry out research. The additional result, however, was to institutionalize the notion that theoretical physics was a separate discipline. Most of these new extraordinary professorships were founded in Prussia. A good example of the way they worked is provided, however, by the circumstances at the University of Strassburg, newly under German administration in the early 1870s, following German victory in the Franco-Prussian War. The first ordinary professor of physics and director of the physics institute there was August Kundt. He soon hired his former student and collaborator Emil Warburg as extraordinary professor of theoretical physics in 1872. While there, Warburg taught theoretical physics as well as collaborating with Kundt on an investigation of the kinetic theory of gases based on Kundt's development of a new method of measuring the velocity of sound. Warburg's responsibility was to conduct the theoretical part of the investigation. Increasingly, the theoretical physicist was the specialist, aiming his teaching at those planning a career in physics, while the experimental physicist lectured to more general audiences.

2.3 The Berlin Physics Institute in about 1877. Prestigious new physics institutes such as this were increasingly important institutional features of German academic science.

new discipline of theoretical physics. His *Kompendium der Theoretischen Physik*, published in two volumes in 1895 and 1896, was a concerted and unprecedented effort to provide a unified view of the new field.

By the final decade of the nineteenth century, the new discipline of theoretical physics was well established in German universities. Gustav Kirchhoff, at Berlin since 1875 as a member of the Prussian Academy, focused his attention more and more on theoretical physics. Hermann von Helmholtz in Berlin both as a director of the Berlin Physics Institute and later as the president of the Physikalisch-Technische Reichsanstalt from 1888 directed his teaching and his research increasingly towards theoretical matters (figure 2.3). After Kirchhoff's death in 1887, Boltzmann was headhunted from Graz to replace him, but declined at the last minute, going instead to a professorship in theoretical physics in Munich, fearing apparently that the Prussians would prove too dour for his taste. Even in the 1890s the position offered to Boltzmann there as an ordinary professor in the new subject was nearly unprecedented. He had the independence of being the director of a state-funded institute with freedom to organize things as he wished. His power was increased when the Austrian government tried to entice him back to Vienna in 1893. The Bavarians responded with an improved salary and the appointment of an assistant. Boltzmann abandoned Munich, however, in 1894 for the professorship in theoretical physics at Vienna and a salary that at 6,000 florins was the highest then paid to an Austrian university professor. It had become a question of national honor that Austria should retain the services of this peripatetic but preeminent theoretical physicist.

This international wrangling over Boltzmann is a good indication of the way that theoretical physics had established itself in German-speaking countries by the end of the nineteenth century. It was a startling achievement considering that the discipline had barely existed less than half a century previously. Half a century later, philosophers of science such as Karl Popper would take it for granted that grand speculation and theorizing was the be-all and end-all of physics, with the experimenter relegated to bottle-washing duties. Theoretically concerned physicists had engineered a spectacular cultural coup by carving out institutional niches for themselves where none had existed before. From being ill-regarded placemen in turn-of-the-century philosophy faculties, physicists—and the new breed of theoretical physicists in particular—had been successful in establishing their discipline at the core of German academic life. The key figures in German theoretical physics—men such as Boltzmann and Helmholtz—were internationally recognized celebrities, not just within their specialist communities but on the broader cultural stage as well. They were recognized as making crucial contributions not just to physics, but to the newly confident and increasingly powerful Germany as well. They helped forged Germany's reputation as a scientific, industrial, and therefore modern state.

## Conclusion

Natural philosophy at the beginning of the nineteenth century looked to many like a potentially revolutionary science. To others it merely looked dangerous. Early nineteenth-century men of science had inherited from the Enlightenment a sense of the ways in which natural philosophy could be used to change the world, to overturn the social order and establish new institutions in its place. For the radicals among them, this sounded like a wonderful idea. To their opponents it was a prospect to be regarded with horror. Depending on one's perspective, natural philosophy could either subvert the proper order of society or reveal in nature what that proper order should be. For those who wanted to put this science to good use, however, the clear conclusion was that scientific institutions needed revolutionizing as well. To make their science matter, they had to find ways of changing these institutions. They had to find ways of changing what it meant to be a man of science. As this chapter shows, different natural philosophers in different countries and locations had a variety of views as to how their practices and their institutions might be transformed. This is not a narrative of continuous and progressive development to a self-evident end.

The state of physics by the end of the nineteenth century was profoundly changed from what it had been at its beginning. At the beginning of the century, natural philosophy was the vocation of a dedicated but tiny band. By the end of the century, new institutions across Europe and America were producing professional physicists in ever increasing numbers. There was no such word as "physicist" in the English vocabulary at the beginning of the century. It was coined by William Whewell to describe what appeared to him to be a new kind of natural philosopher—just as he coined the word "scientist" at about the same time to describe a new breed of practitioner. Its gradual acceptance by the end of the century as the term to describe a particular kind of professional man of science studying nature in a particular fashion was a sign that this new way of doing things had found a secure cultural place for itself. Physicists—and mathematical or theoretical physicists in particular—had managed to secure a vital cultural role for themselves as the ultimate arbiters of what the natural world was like.

# 3

## The Romance of Nature

The worlds of natural philosophy at the beginning of the nine-teenth century were changing rapidly. As we have seen already, many observers regarded natural philosophy as having been (and still being) deeply implicated in the political convulsions that were still sweeping Europe. It seemed to many commentators on both sides of the political fence that the French Revolution and its aftermath were, to at least some degree, the results of Enlightenment philosophies challenging traditional ideas about nature and society. Those political convulsions themselves were responsible for major changes in the institutional structures of European natural philosophy as well. New institutions prolifer-ated and campaigns gathered pace across Europe to reform the moribund structures of ancien régime science. In many ways the social role both of natural philosophy and of its practitioners was up for grabs in such a volatile situation. Not only the insti-tutional context, but the content of science was being quite lit-erally re-formed in the wake of the transformations surrounding the French Revolution. New notions concerning what science should be about—what kind of nature natural philosophers should be searching for—went hand in hand with new visions of what kind of person the natural philosopher should be and how his role in society should be understood and appreciated. Fashioning nature and fashioning the natural philosopher were part of the same process.

One response to the changing contours of natural philosophy and the natural philosophical community was the cultural movement now known as Romanticism. Particularly in the German lands a new generation of natural philosophers tried to identify natural philosophy with the search for a transcendental unity in nature. Reacting against what some of them, at least, perceived as an impoverished Enlightenment insistence on focusing on appearances, Romantic philosophers such as Johann Wolfgang von Goethe, Johann Wilhelm Ritter, and Friedrich Schelling in the German lands and Humphry Davy in England argued for the importance of looking beneath the phenomena in an effort to capture the real underlying unity of nature. Increasingly as well, the search for such unity was held to be the province of a particular kind of individual. Natural philosophy required genius. Only a genius—an inspired individual with access to unique reserves of imagination and intuition—could peer beneath the fractured surface of appearances at the transcendental reality beneath. This novel cult of genius was not unique to natural philosophy. In many ways it was a defining feature of Romanticism across culture in the early nineteenth century. In the arts, literature, and music, as well as natural philosophy, being a genius was very much in vogue.

Laboratories across Europe appeared to be providing more and more grist for the mills of those philosophers intent on discovering unity. Experimenters were finding more and more ways of apparently transforming one kind of natural force into another. The voltaic pile, according to the proponents of the chemical theory at least, seemed to turn chemical forces into electricity. The novel technology of photography seemed to provide a way of using light to produce chemical reactions. Following Hans Christian Oersted's experiments in 1820, it seemed to many that there was some intimate link between electricity and magnetism as well. The most visible technology for turning one kind of force into another was increasingly ubiquitous during the early nineteenth century; the steam engine, according to some natural philosophers, was simply a machine for producing mechanical force from heat. Far from being a transcendental unity, Nature was increasingly seen by some natural philosophers as a laboratory for manipulating and transforming the forces. Conversely, what nature did naturally, the experimenter could now perform in his laboratory. New links could be found in all of this between the natural and the political economy. The way nature's forces were organized provided a powerful model for the way in which human economies should be organized as well.

Laboratory practice revolved more and more about the task of finding new ways of making one kind of force produce another. Such practices could also be used as the foundation of new philosophies of nature—as well as new kinds of politics. A number of natural philosophers before midcentury made efforts to base new systems on the experimenter's ability to transform one kind of force into another. New vocabularies were adopted to express these views concerning the unity of nature. William Robert Grove argued for the correlation of physical forces. Michael Faraday argued for the mutual conversion of forces. James Prescott Joule suggested that force was conserved from one transformation to another. Force was the primary focus in these discussions. Many natural philosophers, particularly in Britain, argued that force should be the fundamental concept of physical science. In their view everyone had an intuitive understanding of what force meant as a result of their own everyday interactions with the world around them—they were continually aware of exerting force, or of having force exerted on them. They argued that this provided a good way of making sure that natural philosophy stayed grounded in the real world of everyday experience. The grand philosophical schemes constructed around the mutual relations of the various forces of nature had another implication too. To claim that the forces of nature were all linked together was to argue as well for the primacy of a natural philosophy that could explain that linkage. In other words, it was a way of reasserting the continued superiority of a general natural philosophy (and a generalist natural philosopher) in the face of what many natural philosophers regarded as a worrying trend towards disciplinary fragmentation and specialization.

By the 1850s and 1860s, however, a new candidate had emerged supreme as the focus at which the physical sciences were united. The second half of the nineteenth century saw the rise of the new science of energy. According to this new science, energy, not force, was the fundamental concept of physics. William Thomson and Peter Guthrie Tait's *Treatise on Natural Philosophy* of 1867 was a quite self-conscious effort to replace Newton's *Principia* as the foundational text of a new kind of natural philosophy. James Clerk Maxwell's work on electromagnetism, culminating in his *Treatise on Electricity and Magnetism* of 1873, was a prime exemplar of the new science's versatility. His powerful synthesis of electricity and magnetism showed how energy could be the new unifying principle of natural philosophy. There was nothing transcendental, however, about this science of energy. Energy, according to this new world picture, was embodied in the ether—a substance that filled all space and operated according to the principles of mechanics, just like a factory engine. Understanding the mechanics of the ether was the holy grail of physics for late nineteenth-century experimenters such as Oliver Lodge.

There was a close link throughout the nineteenth century between the ways in which physics as a discipline was organized and the ways in which physics organized the world. For early nineteenth-century Romantic philosophers, natural philosophy required a particular kind of individual. Apprehending nature's hidden unities required someone with the innate capacity to look beneath the surface of events and see what others could not. Midcentury experimental natural philosophers such as William Robert Grove suggested that the natural philosopher needed to be someone educated to look beyond the limitations of particular disciplinary preoccupations and see the wider picture of the correlation of forces. By the end of the century, proponents of energy physics argued that only those like them, deeply trained in the complexities of mathematical physics, could see the world as it really was. It needed their grasp to comprehend the subtle workings of the ether. Their understanding of that subtle and universal medium gave them the ability to police the sciences—to adjudicate what was and what was not an acceptable way of looking at the world.

## Romantic Science

To modern eyes, the conjunction of science—particularly physics—with Romanticism seems somehow peculiar, or at least surprising. Romanticism evokes images of wild-eyed poets, drugs, and Gothic castles. Physics is taken to be a far more sober affair. That apparent disjunction, however, is very much a product of subsequent history. The Romantic movement, in its origins, was deeply concerned with the problems of constructing a new philosophy of nature and as such quite straightforwardly took the understanding of the physical world as part of its project. In many ways, Romanticism was a response to what its proponents regarded as Enlightenment excesses—an increasing distance from nature that came along with civilization and an increasingly mechanized world picture. In their art, their literature, their poetry, and their science, the Romantics sought to find ways of bridging the gap they saw emerging between modern society and modern individuals and a true understanding of their natural selves and the natural world. What was needed, they argued, was an intuitive understanding of things that transcended mere phenomena and got to the true meaning and unity of nature and man's place in it.

It was a central tenet of Romantic philosophy that nature should be apprehended as a coherent and meaningful whole, rather than as an aggregation of disparate and fragmented phenomena. Romantic philosophers were quite often deeply contemptuous of the Enlightenment tendency towards understanding nature by breaking it down into its constituent parts. The Enlightenment metaphor of the Universe as a clock or a machine whose operations could be understood by taking it apart was the subject for irony, if not of outright mockery. The Romantics instead thought of the Universe as something organic. Like a living thing, the Universe was best approached and appreciated by seeing it as a connected, animated unity. Rather than being taken as separate objects of study, the various phenomena and powers of nature were to be understood as different manifestations of a single underlying and all-embracing cause. The aim of natural philosophy from this perspective was synthesis rather than analysis. Unlike Newton, who insisted that he would not "feign hypotheses" but build his theories on the phenomena, Romantic philosophers insisted that by imaginatively feigning hypotheses they could approach an understanding of a more meaningful reality beneath the surface appearance of things.

The Romantics celebrated a new kind of human being who possessed this capacity to look beneath the surface of things for their true meaning—the genius. A genius, as the term was increasingly used in the late eighteenth and early nineteenth centuries, was possessed of unique capabilities which placed him (and it was invariably a "him") outside conventional limitations. Genius was an innate, rather than a learned, capacity. A man was born with the powers of genius. Quite often the language of possession was used quite literally as well. Genius was an active power that took over a particular individual rather than being something that was under that person's control. There was a strong sense in which it was held that there was a kind of symbiosis between the individual's genius and the natural world that genius allowed him to comprehend. In the fragmented world of early nineteenth-century natural philosophy, this cult of genius played a crucial role in providing a new kind of social place for the man of science. Natural philosophy was represented as a deeply individualistic process in which imagination and inspiration took precedence over collective effort in forging scientific progress.

The monumental German playwright and poet, Johann Wolfgang von Goethe, established many of the parameters of Romanticism in literature, architecture, and the arts, as well as in natural philosophy. Goethe considered his contributions to optics and the science of colors as among his most important works: "I make no claims at all to what I have achieved as a poet. Fine poets were my contemporaries, even finer ones lived before me, and there will be others after me. But that I alone in my century know what is right in the difficult science of colour, for that I give myself some credit, and thus I have a consciousness of superiority to many."[1] Goethe saw his work as a direct attack on Newton's then triumphant theory of optics. In *Zur Farbenlehre* in 1810 he declared that as soon as he saw Newton's "celebrated phaenomena of colours" for himself by looking through a prism, "I immediately said to myself, as if by instinct, that the Newtonian teaching is false."[2] Newton's theory, he argued, was a house of cards built on partial and inconclusive experiments. Newtonians were accused of ignoring observations that contravened their hero's doctrines. At Weimar and in nearby Jena, Goethe built up and participated in a circle of similarly minded philosophers and poets committed to creating an alternative worldview that could be used to challenge, among other things, the Newtonian, mechanistic hegemony in natural philosophy.

The philosopher F. W. J. Schelling arrived in Jena as professor of philosophy in 1798, three years after Prussia had been forced into a peace treaty with revolutionary France; he concurred with Goethe that more was required of natural philosophy than a partial examination of the phenomena. As tutor to a brace of young Saxon aristocrats at the University of Leipzig, Schelling had already published his *Ideen zu einer Philosophie der Natur*, in which he argued forcefully for the necessity of establishing physics on a firm a priori foundation. According to this ambitious master plan for a new *Naturphilosophie*, all of physics was to be deduced from first principles. He elaborated this radical new worldview in 1798 with his *Von der Weltseele*, in which he articulated his grand vision of nature as a living organism. The task of the *Naturphilosoph* was to try to understand the soul of this cosmic being. The end of physics was to be the full comprehension of nature's latent and hidden spirituality, which would, in turn, become a mirror for the further exploration of man's own spiritual nature.

Like Goethe, Schelling was adamant that this higher physics was to be understood quite explicitly as a counter to the sterile mechanics of Newtonianism. He set his vision of the world soul in direct opposition to the world clock that he regarded as encapsulating Newtonianism. Where

---

[1]Quoted in Sepper, "Goethe, Colour, and the Science of Seeing," in A. Cunningham and N. Jardine (eds.), *Romanticism and the Sciences*, 189.

[2]Quoted ibid., 190.

Newton had studied the phenomena of nature as separate cogs and wheels in some grand universal clockwork mechanism, Schelling wanted to see them as mutually interacting members of one single and unified cosmic organism. The key to unlocking the secrets of this world soul was the idea of polarity, or opposites. The heartbeat of the cosmic organism was maintained by the constant interplay and interaction of opposing forces— just like positive and negative electrical forces and the attractive and repulsive powers of magnetism. It was from this grand hypothesis that the rest of physics was to be worked out. In opposition to Newton and his grand dictum of *hypotheses non fingo*—"I do not feign hypotheses"— Schelling insisted that not only was hypothesis permissible, but that it was an essential and integral part of physics. The phenomena of nature were to be understood through his grand hypothesis of polarity—not the other way round. The latest philosophical discoveries were marshaled by Schelling to provide a litany of confirmations and examples of polarity's cosmic role.

Other Romantic thinkers congregated around Jena in about 1800 concurred with many of Schelling's claims concerning the need for a new natural philosophy. Friedrich Schlegel was keen to found his own philosophical system, boasting to a friend that "notebooks on physics I have already, therefore, I think I will soon have a system of physics as well."[3] Schlegel was anxious to bring about a reintegration of science and the literary arts, advising a student that "if you want to penetrate into the very core of physics, have yourself initiated into the mysteries of poetry."[4] The poet and mining engineer Friedrich von Hardenberg (known as Novalis) was also a keen student of natural philosophy, aiming to produce a "scientific bible"[5] that would enumerate and transcend the sum of human knowledge. Another member of the Jena circle, Johann Wilhelm Ritter, was regarded by his interlocutors there as being well on the way to producing the Romantic science they yearned for. Novalis remarked of his friend that "Ritter is indeed searching for the genuine world-soul of nature."[6] A keen follower of Schelling's *Naturphilosophie*, Ritter regarded his task as being to discover the secrets of the *All-Thier* (All-Animal) of the universe. To this end he embarked on an ambitious experimental program to elucidate the notion of polarity in nature.

[3]Quoted in W. Wetzels, "Aspects of Natural Science in German Romanticism," 48.
[4]Quoted ibid., 50.
[5]Quoted ibid.
[6]Quoted ibid., 53.

Ritter, born in Silesia in 1776, had been apprenticed to an apothecary before enrolling as a student of natural philosophy at the University of Jena in 1796. Enthused by Luigi Galvani's and Alessandro Volta's new discoveries in animal electricity as well as by Schelling's claim that electricity was the principle of life in nature, he set out to systematically map the ubiquity of electrical phenomena throughout the natural world. His early work focused on examining the workings of galvanism on organic matter and its relationship to the phenomena and processes of life. He speculated that, at least in principle, electricity might even be used to raise the dead. This, however, was only the first link in a cosmic chain that led back to the fundamental unity of all things: "Where is the sun, where is the atom that would not be part of, that would not belong to this organic universe, not living in any time, containing any time?—Where then is the difference between the parts of an animal, of a plant, of a metal, and of a stone?—Are they not all members of the *cosmic-animal*, of *Nature*?"[7] For many of his Romantic contemporaries in and around Jena and elsewhere, Ritter's experiments were putting the empirical flesh on the bones of Schelling's speculative *Naturphilosophie*.

One avid English reader of these exciting new German speculations in natural philosophy was the poet Samuel Taylor Coleridge. Coleridge immersed himself in writings on natural philosophy while living in the West Country following his departure from Cambridge. As early as 1796 he was planning to visit the German lands in search of philosophical inspiration; after considering Jena he eventually matriculated at the University of Göttingen in 1799. *Naturphilosophie* was central to his efforts during the early years of the nineteenth century to develop a coherent system of thought. Like the Romantic philosophers at Jena, he was convinced that mind was active in nature and that nature itself was an animate, organic whole. In a draft of his seminal *Biographia Literaria* of 1815, he argued that there was a correspondence between the powers of mind and those of nature: "Our Business then is to construct a priori, as in Geometry, intuitively from the progressive Schemes that must necessarily result from such a Power with such Forces, till we arrive at Human Intelligence, and prospectively at whatever excellence of the same power can by human Intelligence be schematized."[8] Nature could, in other words, provide the grounds for the powers of human intelligence that could then be used to

---

[7] Quoted in W. Wetzels, "Johann Wilhelm Ritter: Romantic Physics in Germany," in A. Cunningham and N. Jardine (eds.), *Romanticism and the Sciences*, 203.

[8] Quoted in T. Levere, *Poetry Realized in Nature*, 119.

reflect back on the powers of nature themselves. There was, he argued, a kind of symbiosis between nature and the human mind. It was this symbiosis that made human understanding of nature possible in the first place. Like the Germans, he condemned English science for its mechanistic sterility, contrasted to what he regarded as the dynamism of the new German school.

Coleridge's close friend Humphry Davy was another English advocate of Romantic science. Davy, who like Ritter had started his career as an apothecary's apprentice before joining the radical chemist Thomas Beddoes at his famous (or infamous) Pneumatic Institute in Bristol, also like his German contemporary founded his early reputation on the exciting new science of galvanism. Closely associated with the poets Robert Southey and William Wordsworth as well as Coleridge, during his sojourn in the West Country, Davy dabbled in poetry himself while engaged in his chemical and electrical experiments. Like the German Romantics, Davy was simultaneously obsessed with the powers of nature and the powers of human genius (of which he regarded himself as being a preeminent example). The discovery of nature was taken to be a means to self-discovery in Davy's scheme of things—sometimes quite literally so. In early experiments on the newly discovered gas nitrous oxide, Davy experimented with breathing the new air as a means of investigating his own mind and wrote rapturous poetry of the resulting experiences.

After leaving the Pneumatic Institute to take up a post lecturing at the new Royal Institution in London, Davy seemed, at least to some of his erstwhile collaborators, to have abandoned his radical and Romantic roots for the carrot of metropolitan success. His experimental researches throughout the 1800s and 1810s, however, continued to focus on uncovering the active powers of nature and establishing their fundamental unity. For Davy, the various powers of chemical affinity, electricity, heat, and light were all to be regarded as different manifestations of one underlying and all-embracing active principle. The natural philosopher's task, as he saw it, was to make these different powers available for mankind's material benefit as well as their spiritual uplifting. The philosopher's genius that led to an intuitive understanding of nature also conferred power over it: "By means of this science man has employed almost all the substances in nature either for the satisfaction of his wants or the gratification of his luxuries. Not content with what is found upon the surface of the earth, he has penetrated into her bosom, and has even searched the bottom of the ocean for the purpose of allaying the restlessness of his desires, or of extending and increasing his power."[9] This utilitarian twist on Romantic philosophy was particularly apposite to his position at the Benthamite-inspired Royal Institution and later in the 1820s as president of a reforming Royal Society.

Romantic science in the early nineteenth century was a loose collection of practices and ideas rather than a coherent school or system. Even Romantics like Novalis, for example, could be scornful of some aspects of Schelling's *Naturphilosophie*, dismissing it as little more than the "boasting of a fanciful mind."[10] In some respects, maybe, a philosophy that placed so much emphasis on the role of individual inspiration in the discovery of nature was not aimed in any case at producing a collective research program. Future generations of natural philosophers, particularly in the German lands, were scathing in their condemnation of the metaphysical excesses of their predecessors. Romanticism provided, however, some practical and intellectual tools for self-fashioning and disciplinary fashioning in the opening decades of the nineteenth century. In the face of massive social and intellectual upheaval at the turn of the century it helped construct a new vision of nature as a unified, organic whole, which in turn provided a resource both for constructing new scientific disciplines—including physics—and for constructing a new image of the natural philosopher and his place in society.

## The World's Laboratory

New laboratories sprang up all over Europe during the first few decades of the nineteenth century. As we shall see, these early nineteenth-century laboratories differed from their predecessors in kind as well as number. They were starting to be regarded as spaces for research in their own right rather than merely as adjuncts to the lecture theater. Experiment increasingly looked like the key to unlocking nature's secrets and putting them to the service of humankind. Furthermore, what more and more experimenters in these new laboratories seemed intent on doing was finding ways of turning one kind of force into another. To Romantic natural philosophers, these proliferating instances of the apparent conversion of one kind of force to another seemed to be powerful confirmatory evidence

---

[9] Quoted in Lawrence, "The Power and the Glory: Humphry Davy and Romanticism," in A. Cunningham & N. Jardine (eds.), *Romanticism and the Sciences*, 221.

[10] Quoted in W. Wetzels, "Aspects of Natural Science in German Romanticism," 49.

of the underlying unity of all the forces of nature. They were mutually convertible because they were all simply different manifestations of the one underlying power. More pragmatically perhaps, others saw economic potential in this ability to turn one force into another. Laboratory experiments might prove to be the means of delivering inexhaustible new sources of exploitable power. Not only the powers of wind and water (or humans and animals) might be made to do useful work, but the forces of chemistry, electricity, heat, and light might all be made available.

There was an underlying metaphysics behind this kind of economic interest as well, however. By the end of the eighteenth century many natural philosophers agreed that nature had its own economy and that the political economy mirrored (or at least ought to mirror) the natural one. Putting the resources of the one at the disposal of the other seemed appropriate. Some argued that the natural economy was designed to be placed at the disposal of humanity in any case. The geologist James Hutton argued in his *Dissertations on Different Subjects in Natural Philosophy* of 1792 that nature should be regarded as a self-regenerating system of active powers with the Sun as the source of regenerative power. Adam Walker in his *System of Familiar Philosophy* of 1799 maintained that light, fire, and electricity were simply different manifestations of a single principle of repulsion and that nature's economy was maintained as a perpetual balance between this principle of repulsion and the attractive principle of gravitation. The radical natural philosopher and political writer Joseph Priestley regarded phlogiston—the active principle of combustion—as underlying all of nature's economy. Humphry Davy in his *Essay on Heat, Light and the Combinations of Light* (1799) asserted that electricity, light, and chemical action were different manifestations of the same thing.

The flamboyant American émigré and Royalist, Benjamin Thompson, ennobled as Count Rumford by the elector of Bavaria, was a keen advocate of the utility of the experimental sciences. As such he was to play a crucial role in promoting utilitarian scientific enterprises at the beginning of the nineteenth century. He was a key figure in the establishment of the Royal Institution in London in 1799, having helped convince a group of reforming landowners of the prospects for putting chemistry and natural philosophy to work in improving agriculture. Rumford was a staunch opponent of the caloric theory of heat—the notion that heat was an imponderable fluid. He carried out experiments on cannon boring, showing how friction produced heat. He argued that the amount of heat produced during the process was indefinite and suggested therefore that heat could not be a fluid but was instead the result of particles in motion. Humphry Davy, professor of chemistry at Rumford's Royal Institution within a few years of its foundation, was similarly an opponent of the caloric theory, then popular in Revolutionary France. Its last great advocate there had been Lavoisier, whose wife Rumford had married following her former husband's death at the guillotine. Like Rumford, Davy carried out some confirmatory experiments, in this case on the melting of ice due to friction. By showing that blocks of ice rubbed together melted, Davy suggested that the heat produced by the process was the result of friction and that heat was simply a kind of motion.

As we shall see in considerably more detail in a later chapter, heat and its relationship to motion was a matter of particular concern to early nineteenth-century experimenters. With the Industrial Revolution in full swing, the steam engine in particular was taken to be the preeminent example of a device that turned one kind of force—heat—into another—motion. It also seemed to be a machine designed to make the power tied up in the "vast storehouse of nature" available to humankind. The young Mancunian brewer's son, James Prescott Joule, in the 1840s devoted considerable experimental effort to improving the efficiency of steam engines. That was what underlay his efforts to experimentally determine what he called the "mechanical equivalent of heat." He had already made a name for himself carrying out experiments assessing the efficiency of electromagnetic engines. His famous paddle wheel experiment was a way of showing that motion and heat were literally interchangeable and of providing the data that allowed him to calculate the exact rate of exchange.

Many of the early nineteenth century's laboratory conversion processes were offshoots of the voltaic battery—the dramatic new technology that was the culmination of Alessandro Volta's dispute with Luigi Galvani concerning the origins of animal electricity. Ironically, however, Volta himself did not consider the voltaic battery a conversion device. He regarded the source of electricity in the battery as the simple contact of the metals. Only for his opponents who like Davy held that the electricity was produced from chemical affinity, or Galvani himself, who held that electricity was identical with the nervous force of the body, was the battery the instrument of the conversion of one kind of force or power into another. Galvani's nephew Giovanni Aldini used the battery to spectacular effect to confirm his uncle's claims concerning animal electricity. Experimenting on the bodies of dead animals and of executed felons, he could show how electricity seemingly reproduced the appearance of life in otherwise inanimate corpses (figure 3.1). The Royal Humane Society in England speculated that Aldini's experiments might prove to

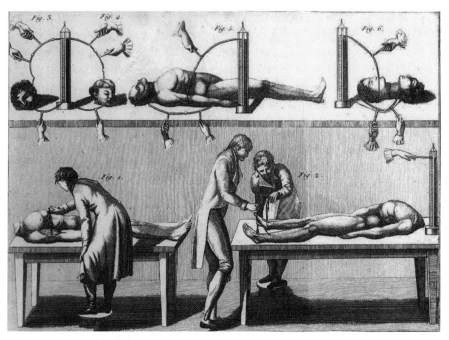

3.1 Electrical experiments on recently executed criminals from Giovanni Aldini, *Treatise on Galvanism* (1803).

be the key to restoring life to the prematurely dead. The Scottish chemist Andrew Ure made similar claims when he carried out electrical experiments on an executed murderer in Glasgow in 1818.

The body, human and animal, was increasingly a focus for experimentation on relationships among the powers and forces of nature. Ritter, coming as he did from a Romantic perspective that regarded the entire Universe as a single connected organism, was deeply involved in experimental work on the relationship between electricity in particular and the nervous powers responsible for animating animal and human bodies. His *Das elektrische System der Körper*, published in 1805, was an ambitious effort to construct a cosmic electrical system in which electricity was held up as the basic organizing principle of all animate and inanimate matter. To this end Ritter experimented copiously on the actions of electricity on the body, its effects on the growth of plants, and so forth. Much of his work was dismissed as fanciful by contemporaries and successors anxious to dissociate themselves from the *Naturphilosophie* that informed his experimental researches. A few decades later the Italian natural philosopher Carlo Matteucci embarked on his own experimental program to investigate the electrical properties of living tissue and the question of the relationship between electricity and the nervous force. He produced a long series of papers through the 1840s, translated into English and published in the *Philosophical Transactions* of the Royal Society, on experimental efforts to measure electrical currents in the body.

Hans Christian Oersted, a friend of Ritter's and like him an admirer of Schelling's *Naturphilosophie*, carried out experimental work aimed at finding links among the different powers of nature as well. Oersted was particularly fascinated by the possibility of establishing a relationship between electricity and magnetism. It seemed to many experimenters that there must be such a connection, if only because of the many analogies they perceived between electricity and magnetism. In 1820 Oersted, by now professor of physics at the University of Copenhagen, succeeded in finding the long sought-for link. He found that when a magnetized needle was suspended near a copper wire carrying a flow of electricity from a voltaic battery, the needle deflected (figure 3.2). Oersted interpreted his findings in terms of an "electric conflict" surrounding the wire and interacting with the needle's magnetism. He argued that heat and light should be regarded as the result of such electric conflicts as well. Oersted's work was enthusiastically reproduced at the Royal Institution in London, where the young Michael Faraday succeeded in making an electric wire rotate around a magnet. A decade later he succeeded in producing electricity from magnetism as well. To those looking for unity in nature, it appeared that more and more links in the chain were being forged all the time.

Another link was forged in Berlin. Thomas Johann Seebeck, an independently wealthy experimental enthusiast and devotee of Goethe's theory of colors, had been fascinated by the possibility of finding connections between heat and light since 1806. He was particularly interested in the relationship between heat and the colors of the solar spectrum. Inspired by Oersted's discovery, he set out on an experimental examination of the connections between electricity, magnetism, and heat. His aim was to produce magnetic phenomena by heat. Instead, he found a way of producing electricity from heat. He found that if he constructed a circuit partly of copper and partly of bismuth and heated one of the junctions where the two metals joined, a current was registered by a magnetized needle suspended nearby. William Sturgeon in England used Seebeck's discovery to good effect, showing how a spherical cage made of the wire of the two metals could be made to rotate about a central magnet when the junctions were heated. This was the solution to why the Earth

**3.2** A nineteenth-century artist's impression of Hans Christian Oersted's 1820 observation that a magnetized needle held near a wire carrying a current of electricity was deflected.

rotated on its axis, according to Sturgeon, a self-taught natural philosopher and former artilleryman based in Woolwich near London and the inventor of the electromagnet. The currents produced by the Sun's heat made the Earth rotate about the central magnetic core—as well as producing spectacular electrical storms in the tropics. A dozen years later, in 1834, the Frenchman Jean Charles Peltier added another link, showing that electricity could absorb heat as well, reversing Seebeck's experiment and using electricity to produce cold.

Heat and light were also coming to be regarded by some investigators as different manifestations of the same thing. The astronomer William Herschel, internationally fêted for his discovery of the planet Uranus, had carried out experiments on the temperature of the solar spectrum in the 1790s and had shown that temperature varied from one end of the spectrum to the other—the red end of the spectrum being hotter than the violet one. From this Herschel proceeded in 1800 to the discovery of infrared light beyond the red portion of the spectrum. Inspired by Herschel's discovery and anxious to preserve the spectrum's symmetry, Ritter postulated that there must be yet another invisible part of the spectrum at the other, violet end. In 1801, Ritter succeeded in showing that something emanating from just beyond the violet end of the spectrum darkened a silver chloride compound just as ordinary light did. He had found the ultraviolet part of the spectrum. The Italian Macedonio Melloni's experiments on radiant heat during the 1830s—showing that it could be made to manifest physical properties similar to those of light—were also widely regarded as making the identity of light and heat secure.

Another new and rapidly developing technology of the early nineteenth century seemed to demonstrate conclusively the relationship between the powers of light and chemical affinity. Experimenters had long known that light had an effect on certain chemical compounds (like the silver chloride in Ritter's experiments on ultraviolet light). The development of photography, however, had the effect of making this relationship graphically visible. Light shining onto surfaces specially treated with certain chemicals could produce stunning images of real-life scenes, objects, and even people. Louis Jacques Mandé Daguerre in France and William Henry Fox Talbot in England both developed successful techniques for creating and preserving these impressive and novel images. Fascinated commentators regarded the new technology as an unprecedented example of nature's powers being put at the service of humankind. As the experimental natural philosopher William Robert Grove joked, the day would soon come when "instead of a plate being inscribed, as 'drawn by Landseer, and engraved by Cousins,' it would be 'drawn by Light, and Engraved by Electricity!'"[11] Michael Faraday enthused over Talbot's

[11] W. R. Grove, "On a Voltaic Process for Etching Daguerreotype Plates," *Philosophical Magazine*, 1842, 20: 24.

"photogenic drawings," exclaiming that "what man may do, now that Dame Nature has become his drawing mistress, it is impossible to predict."[12]

Many commentators regarded these proliferating examples of the apparent conversion or interchangeability of one force for another as evidence of at least some kind of underlying unity, though just what kind of unity it might be was a subject for debate. In England certainly, many took a similar perspective to that of Humphry Davy, who had argued that all these manifestations of force were to be regarded as different ways in which nature's powers had been made available for humanity's benefit. Charles Babbage in his *Economy of Machinery and Manufactures* in 1832, providing his perspective on the sources of economic wealth and progress, started with an overview of the natural powers that produced useful work and the machines that could be used to make those powers available for exploitation. Central to that analysis was Babbage's insistence that all these various machines should be regarded as technologies that transformed power rather than creating it. Windmills, waterwheels, steam engines, and so on were properly seen as machines that converted the powers of nature into a form that could be made useful. Other commentators on political economy concurred. Nature presented a wealth of interrelated powers. The task of natural philosophers and experimenters was increasingly plausibly regarded as finding practical ways to put those powers to work.

There was a big difference between the grand metaphysical accounts of nature's unity that prevailed in the late eighteenth century—and in many ways reached their apogee with Schelling's *Naturphilosophie*—and the proliferating force transformations of the first few decades of the nineteenth century. The force transformations resulted from practical laboratory technologies. The experiments that produced these transformations often ended up looking more like machines designed to convert one kind of power into another than like indications of underlying organic unity. New analyses of political economy looked to machinery as a way of explaining progress. As human labor (itself from this perspective the result of yet another force transformation) was increasingly replaced by more and more productive machinery, economic progress would be indefinite. New technologies like batteries, electromagnetic engines, thermoelectric couples, and even cameras could be slotted conveniently into economic stories that pointed to machines designed to maximize the efficient production of power from natural resources as the source of new

[12]Quoted in I. R. Morus, *Frankenstein's Children*, 175.

social and economic progress and of wealth. They provided new ways of forging links between the progress of natural and political economies.

## Correlating the Forces

Several natural philosophers, particularly in Britain, turned to the new laboratory experiments developed during the first half of the century to provide the building blocks for ambitious new accounts of the natural economy. At the very least, practical examples of how the unity of nature could be made tangible and useful provided vivid illustrations of metaphysical principles. Audiences at William Robert Grove's lectures at the London Institution or at one of Michael Faraday's Friday evening discourses at the Royal Institution during the 1840s could literally see forces being transformed one to another. It is no accident that the first public utterances by both these natural philosophers concerning the interconvertibility of natural powers were made in lectures before a popular audience. As well as spectacular demonstrations, however, experiments like these provided flesh for the bones of new accounts of nature and the progress of natural laws. In an increasingly hardnosed and utilitarian culture they also provided solid examples of why such accounts of the natural economy really mattered. They showed how natural philosophical principles could be used to make nature's work useful. They were a good way, therefore, of explaining to the Victorian public how natural philosophers themselves were useful and productive as well.

Grove, a Welshman from Glamorganshire, had been appointed professor of experimental philosophy at the London Institution in 1841. That institution, founded in 1805 by a clique of City businessmen as a rival to the more aristocratic Royal Institution on Albemarle Street, was itself devoted to solidifying the link between science and commerce. Charles Butler declared at the London Institution's inauguration ceremony that science and commerce combined would "record the heavens, delve the depths of the earth, and fill every climate that encourages them with industry, energy, wealth, honour, and happiness." When they are separated, "Science loses almost all her Utility; Commerce almost all her dignity."[13] Grove, a graduate of Brasenose College Oxford who had trained for the bar at Lincoln's Inn, had made his scientific reputation with his work on the construction of powerful and long-lasting electric batteries. He would have agreed with Butler's sentiments. One of his first tasks

[13] C. Butler, *Inaugural Oration* (London, 1816), 40.

following his appointment as professor at the London Institution was to deliver a public lecture on the progress of the physical sciences since the institution's founding. He took advantage of the opportunity to deliver his own encomium on the importance of linking science and industry.

Grove's lecture placed recent developments in physics within the context of a progressive ideology that saw developments in science and society as necessarily going hand in hand. The discoveries he would discuss, he claimed, had already "wrought epochal changes in our knowledge, and will work gradual changes in our political history."[14] His lecture provided his audience with a tour of the latest science, from the steam engine ("the grandest mastery of mind over matter") through electricity and magnetism to photography ("the most remarkable discovery of modern times"), with constant emphasis on their utility and contribution to social progress. As Grove put it, "The student who in his closet successfully interrogates Nature, not only gives to man new physical knowledge, but works an indelible change in his moral destinies."[15] Running through the lecture was the view that the recognition that nature's powers were interconnected lay at the root of recent discoveries and their social benefits: "Light, Heat, Electricity, Magnetism, Motion, and Chemical-affinity, are all convertible material affections; assuming either as the cause, one of the others will be the effect."[16] It was humankind's capacity to manipulate these relationships that resulted in scientific, social, and economic progress: "Why is England a great nation? Is it because her sons are brave? No, for so are the savage denizens of Polynesia: She is great because their bravery is fortified by discipline, and discipline is the offshoot of Science. Why is England a great nation? She is great because she excels in Agriculture, in Manufactures, in Commerce. What is Agriculture without Chemistry? What Manufactures without Mechanics? What Commerce without Navigation? What Navigation without Astronomy?"[17]

Grove soon coined a new phrase for his conviction that nature's powers were mutually convertible: the correlation of physical forces. In a series of popular lectures at the London Institution, published as an essay when Grove left his post there in 1846, correlation was used as the organizing principle around which he arranged his survey of the physical sciences. His position was "that the various imponderable agencies, or

---

[14]W. R. Grove, *On the Progress of the Physical Sciences* (London, 1842), 6.

[15]Ibid., 30.

[16]Ibid., 30–31.

[17]Ibid., 37.

the affections of matter which constitute the main objects of experimental physics, viz. Heat, Light, Electricity, Magnetism, Chemical Affinity, and Motion are all Correlative, or have a reciprocal dependence. That neither taken abstractedly can be said to be the essential or proximate cause of the others, but that either may, as a force, produce or be convertible into the other, thus heat may mediately or immediately produce electricity, electricity may produce heat; and so of the rest."[18] Correlation played a role as a rhetorical device for Grove, providing him with a narrative structure that held together the experimental displays that made up his lectures, while at the same time, of course, the experiments' role became that of making correlation visible (figure 3.3).

The theatricality of correlation is evident in one of Grove's examples, designed to show the production of all the other modes of force from light. In this experiment a Daguerreotype plate was placed in a glass-fronted box filled with water, along with a grid of silver wire connected to the plate to form a circuit along with a galvanometer and a Breuget helix. When light fell on the plate following the removal of a shutter covering the glass front, the galvanometer needles twitched and the Breuget helix expanded. As Grove explained, "[T]hus, Light being the initiating force, we get chemical action on the plate, electricity circulating through the wires, magnetism on the [galvanometer] coil, heat in the helix and motion in the needles."[19] This kind of display of correlation was important in Grove's own experimental work as well. For him, the main significance of the gas battery (the ancestor of the modern fuel cell), which he invented in 1842, was that it represented "such a beautiful instance of the correlation of natural forces."[20] In the gas battery, electricity was produced through a process that combined oxygen and hydrogen to produce water. The electricity produced could then be used to decompose the water into its constituent elements once more. Grove saw this as the ultimate example of correlation in action.

Like Grove, Faraday first made his thoughts concerning the interrelationships of nature's powers public in a lecture, this time to one of the Royal Institution's popular Friday evening discourses. Faraday had been instrumental in first establishing the discourses in the 1820s as a forum for showcasing the latest scientific discoveries and spectacular new inventions. By the 1840s they were immensely popular, Faraday himself

[18] W. R. Grove, *On the Correlation of Physical Forces* (London, 1846), 7–8.
[19] Ibid., 28.
[20] W. R. Grove, *Literary Gazette*, 1842, 26: 833.

3.3 The frontispiece of Henry M. Noad's *Lectures on Electricity*
(1844) showing an idealized (if messy) electrician's
workshop. In the foreground is an Armstrong hydro-
electric machine, and a Wheatstone and Cooke five-needle
telegraph hangs on the wall behind it. Various items of
electricians' apparatus such as induction coils and
batteries are scattered around. On the right, on top of the
arch, is Grove's gas battery, illustrating the correlation of
physical forces.

drawing enthusiastic crowds of hundreds when he performed. This was
the context in which Faraday first made public his speculations concerning the interconvertibility of the forces and the relationship between force
and matter. He kicked off the 1844 season of Friday evening discourses
on 19 January with a lecture entitled "A Speculation Concerning Electric

Conduction and the Nature of Matter." He attacked conventional notions of matter as solid particles, suggesting instead that matter should be regarded as being a manifestation of force. His argument was that since matter is only recognized through the forces it exerts, matter and force should be regarded as in some sense identical. Rather than visualizing the material world as made up of solid particles in space, Faraday saw it as a plenum through which forces acted on each other.

Faraday's claims concerning the conservation of force and of matter was in many ways a theological imperative. God had created a certain, finite amount of matter and of force. Since this was created by God, it could in no way be destroyed by any other agency than God's. Force was therefore conserved in any interaction. Faraday was less convinced, however, that forces were interconvertible. He was clear that they were interrelated—much of his experimental work was devoted to demonstrating such relationships, as with his work on electromagnetism and magneto-optics and his efforts to find a relationship between electricity and gravitation—but was unconvinced that this interrelationship could produce actual conversion from one kind of force to another. In his referee's report to the Royal Society on James Prescott Joule's experiments on the mechanical equivalent of heat, Faraday recommended that Joule modify his conclusions for precisely these reasons. Joule wanted to argue that his experiments showed that heat and motion really were mutually convertible—that a particular amount of motion could be turned into a particular amount of heat and vice versa. Faraday disagreed. He argued that all the experiments showed was that a certain amount of motion always resulted in the same amount of heat and that nothing further could legitimately be inferred from the experiments. Joule had to modify his conclusions to suit Faraday's objections.

Like Faraday, however, Joule took the conservation of force to be a fundamental theological principle. The difference between their views is perhaps best encapsulated by the observation that Faraday believed in the conservation of forces while Joule believed in the conservation of force. Joule took it as axiomatic that nothing that God had created could be destroyed by man and that force, therefore, simply had to be conserved in any interaction. It was a fundamental assumption rather than an experimentally derived generalization. Coming from provincial Manchester rather than the metropolis, Joule had to find alternatives to the prestigious London institutions in order to make his voice heard. By 1842 he had been elected a member of the local Manchester Literary & Philosophical Society, having delivered a paper there a few months previously entitled "The Electrical Origin of the Heat of Combustion." He also performed before the British Association for the Advancement of Science when they gathered in Manchester later that year and on other occasions later in the 1840s, notably in 1847 when his experiments caught the attention of the young William Thomson. The most comprehensive presentation of his views on the conservation of force was made, however, to a local and provincial audience gathered for a public lecture at St. Anne's Church School in Manchester in 1847.

Illustrating his performance with such force transformations as experiments with voltaic batteries and electromagnetic engines, Joule aimed to convince his audience of the reality of conservation and conversion processes in nature. It was an ambitious program. He needed to demonstrate conclusively that "the phenomena of nature, whether mechanical, chemical or vital, consist almost entirely in a continual conversion of attraction through space, living force and heat into one another. Thus it is that order is maintained in the universe—nothing is deranged, nothing is ever lost, but the entire machinery, complicated as it is, works smoothly and harmoniously."[21] Any apparent loss of living force (as he translated the eighteenth-century Latin mathematical term, *vis viva*) was simply the result of the conversion of that force into another form according to a strict principle of equivalence. As in the case of Grove and his principle of correlation, Joule was using his notion of the conservation of force to place his kind of physics right at the center of natural philosophy. If he was right, then all nature and all natural processes were governed by the principle to which he laid claim.

Grand and overarching accounts of natural philosophy like this were increasingly popular during the 1830s and 1840s in England. John Herschel's *Preliminary Discourse on the Study of Natural Philosophy*, published in 1830 as the introductory volume of Dionysius Lardner's Cabinet of Natural Philosophy, laid out a synthetic framework for the sciences. The same could be said for William Whewell's magisterial *History of the Inductive Sciences* and its companion *Philosophy of the Inductive Sciences*. Mary Somerville's *Connexion of the Physical Sciences* published in 1834 had a similar agenda. Such texts were bestsellers for their time, going through numerous editions and revisions. They laid out a range of visions of the sciences as a unified whole at just the historical moment when natural philosophy seemed to be fragmenting into a myriad specialist disciplines. The conjunction is hardly coincidental. Texts like these could

[21]Quoted in C. Smith, *The Science of Energy*, 72.

be used to try and hold the edifice of science together when it appeared to be in danger of breaking up. Crucially, they offered a range of new accounts as to what kind of role in culture the natural philosopher might hope to play.

This spate of mid-nineteenth-century syntheses of the physical sciences were exercises in marking out new territory. All these scientists—a word coined by William Whewell in 1832—were trying to redefine the field of their science, as well as the cultural role of science and scientists. New accounts of the correlation, conservation, and conversion of forces served as means of redefining and restating the importance of physics, both as an intellectual exercise and as an economic one. Syntheses such as these, drawing on the latest experiments, showed how nature could be put to work in earnest. The natural philosopher's role, on this showing, was to manage the process of making nature's powers part of the workforce. The laboratory technology that was used to make correlation or conversion visible—the voltaic batteries, electromagnetic devices, and photographic apparatus—could be made to join the steam engine on the factory floor. In these new articulations of the ways in which nature was hooked together, laboratory apparatus could be used to forge secure links between progress in nature and progress in society. From this perspective, the implication was that natural philosophers had a central role to play, not only in uncovering nature's secrets but in placing them firmly at the center of the cultural stage.

### Energy's Empire

A new word appeared in physics at about midcentury: energy. The word was bound up with a new doctrine as well: the conservation of energy. By the end of the nineteenth century all of physics and much of the other sciences as well revolved around this new concept. Its early proponents— physicists such as William Thomson (later Lord Kelvin) and James Clerk Maxwell—argued vociferously that physics needed to be reorganized around this fresh idea if it was to be placed on a secure footing. Opponents, particularly the eminent astronomer Sir John Herschel, disagreed strongly. They argued that the old concept of force should retain its preeminence since everyone had an instinctive appreciation of its meaning from their everyday experiences. Energy in comparison was a chimera, a theoretical construct with no tangible expression in the real world. By the end of the century, by contrast, many physicists would have argued that energy *was* the real world. For its promoters, energy was the hidden link that bound the disparate phenomena of nature together. Energy, not force, was the quantity conserved during the transformation of electricity into heat, or heat into motion. They aimed to use the new concept to forge a powerful new science, tailored for the modern industrial age.

The promoters of this new science of energy were quite self-conscious in their efforts to revolutionize not just physics but the whole of natural philosophy. They aimed at an achievement comparable to Newton's in providing a new and secure foundation for their science. A good indication of what was at stake was the recurrent bickering over who could be credited with the discovery of the grand principle of the conservation of energy. William Thomson and his irascible close ally Peter Guthrie Tait, professor first at Queen's College Belfast and then at Edinburgh, hailed James Prescott Joule as the discoverer. The materialist John Tyndall, friend of T. H. Huxley and admirer of German science, pointed to the obscure German physician Robert Mayer instead, unwilling to give the laurels to the devout Anglican Joule. William Robert Grove pointed to his own *The Correlation of Physical Forces* as an early enunciation of the principle. For the energy physicists, however, Grove's claim was "humbug" and the man himself though "not a bad fellow" was "woefully loose and unscientific."[22] For them, Grove was one of the old guard, lacking in the rigorous mathematical apprenticeship needed to appreciate the new theory in all its glory.

The doctrine of conservation soon acquired its bible with the publication of the ambitious *Treatise on Natural Philosophy*, coauthored by William Thomson and Peter Guthrie Tait between 1862 and 1867. The aim of the book in many ways was to make energy real—to bring the abstractions of the Cambridge mathematical Tripos down to Earth. As one reviewer put it: "The world of which they give the Natural Philosophy is not the abstract world of Cambridge examination papers—in which matter is perfectly homogenous, pulleys perfectly smooth, strings perfectly elastic, liquids perfectly compressible—but it is the concrete world of the senses, which approximates to, but always falls short of the mathematical as of the poetical imagination."[23] James Clerk Maxwell agreed: "The two northern wizards were the first who, without compunction or dread, uttered in their mother tongue the true and proper names of those dynamical concepts which the magicians of old were wont to invoke only

[22] P. G. Tait to W. Thomson, 2 December 1862, quoted ibid., 176.
[23] Quoted ibid., 192.

by the aid of muttered symbols and inarticulate equations."[24] It was an answer to the complaint that the new energy physics was too abstract, too divorced from mundane reality. The text provided a comprehensive articulation of the doctrine of the conservation of energy, insisting on and illustrating its universal application throughout the physical world and on its foundational role in physics. It also invoked a highly distinguished pedigree for the new idea.

Energy's precursor was no less a sage than the illustrious Isaac Newton. In their preface, Thomson and Tait (or T and T', as they facetiously referred to each other in private correspondence) announced this heritage proudly: "One object which we have constantly kept in view is the grand principle of the *Conservation of Energy*. According to modern experimental results, especially those of JOULE, Energy is as real and indestructible as Matter. It is satisfactory to find that NEWTON anticipated, so far as the state of experimental science in his time permitted him, this magnificent modern generalization."[25] They were themselves, they announced, "Restorers" rather than "Innovators." They presented the conservation of energy as placing physics back onto the true path first blazed by Newton himself. This was, in part, a campaign to guarantee for energy an impeccably British ancestry against the claims of German interlopers such as Mayer and Helmholtz. In the same way, the insistence that Joule, in particular, should be recognized as the discoverer was, as much as it was aimed to satisfy English amour propre, also aimed at underlining the claim that the conservation of energy was a firmly empirical discovery, solidly based on experimental evidence rather than being derived from airy and abstract speculation.

In popular lectures, magazine articles, and books, Thomson and Tait, along with allies such as Balfour Stewart, professor of natural philosophy at Manchester's Owens College, set out to proselytize for the new doctrine. They wanted to show fellow scientists and the broader public that the conservation of energy was far more than just a narrow scientific principle but that it applied to the whole of natural philosophy and beyond. It was an idea that could explain everything from the movements of the stars and planets down to the mechanism of life itself. As Stewart put it in his *The Conservation of Energy* (1873), the Universe could be regarded "in the light of a vast physical machine" and "the laws of

[24]J. C. Maxwell, "Thomson and Tait's Natural Philosophy," *Nature*, 1879, 20: 215.
[25]W. Thomson and P. G. Tait, *Treatise on Natural Philosophy* (Oxford, 1867), vi.

energy as the laws of working of this machine."[26] His popular introduction, published as part of the International Scientific Series, took the reader through energy and its transformations from the energy of a rifle bullet fired from a gun to the infinitesimal chemical reactions taking place in the human or animal body. In *The Unseen Universe or Physical Speculations on a Future State* (1875), Stewart and Tait proclaimed the complete conformity of the new physics to Christian orthodoxy. They went on to show that even "immortality is strictly in accordance with the principle of Continuity (rightly viewed); that principle which has been the guide of all modern scientific advance."[27]

The new doctrine of energy was also at the heart of James Clerk Maxwell's attempt to construct a comprehensive theory of electromagnetism from Faraday's experimental researches and scattered speculations. Born into a genteel Edinburgh family in 1831, Maxwell was first taught natural philosophy by James Forbes at the university there while imbibing metaphysics from Sir William Hamilton. In 1850 he went to Cambridge to study for the mathematical Tripos, first at St. Peter's and then at Trinity College. He made quite an impression, graduating as second wrangler and joint Smith's prizeman in 1854 and obtaining a college fellowship at Trinity a year later. (The Smith's Prize for mathematical proficiency was established by the bequest of Robert Smith, a former master of Trinity College, on his death in 1768.) Cambridge's rigorous training in mathematical analysis provided him with the skills he would need as an ambitious young physicist eager to make his mark on the new science of energy. Maxwell was encouraged by Thomson to familiarize himself with Faraday's researches; his paper "On Faraday's Lines of Force," read to the Cambridge Philosophical Society in 1855 and published in their *Transactions* a year later, was his first essay into the field. He returned to the topic some years later, publishing "On Physical Lines of Force" in the *Philosophical Magazine* between 1861 and 1862. The papers were part of a concerted effort to turn Faraday's speculations about lines of force in space into something more concrete.

Maxwell appropriated Faraday's speculations to provide a solid pedigree for his mathematics of energy. "Nothing is clearer," he told Faraday, "than your description of all sources of force keeping up a state of energy in all that surrounds them . . . You seem to see the lines of force curving round obstacles and driving plump at conductors and swerving towards

---

[26]B. Stewart, *The Conservation of Energy* (London, 1873), v.

[27]B. Stewart and P. G. Tait, *The Unseen Universe* (London, 1875), 28.

certain direction in crystals, and carrying with them everywhere the same amount of attractive power spread wider or denser as the lines widen or contract."[28] As far as Maxwell was concerned, those lines of force were real. His task was to find the mathematics that described their behavior. In "On Physical Lines of Force" he elaborated a complex mechanical model of molecular vortices and idle wheels to represent his theory. This was the kind of medium that his mathematics described—the ether. It might not be the model that existed in reality, but it was "a mode of connexion which is mechanically conceivable, and easily investigated."[29] In "A Dynamical Theory of the Electromagnetic Field," published in the Royal Society's *Philosophical Transactions* in 1865, Maxwell refined the theory yet again drawing on the latest mathematics to flesh out his understanding of the ether—the space-filling medium in which electromagnetic energy was stored and through which it traveled.

Maxwell's electromagnetic theorizing culminated with the *Treatise on Electricity and Magnetism* (1873), published just two years after he had been appointed first Cavendish Professor of Physics at the University of Cambridge (figure 3.4). Like his energetic allies Thomson and Tait, he was trying to build the foundations of a comprehensive new science based on the concept of energy. The treatise brought together and elaborated the fruits of his earlier papers, putting the flesh on the bones of the electromagnetic ether. As Maxwell had already explained in his "Dynamical Theory," "In speaking of the Energy of the field . . . I mean to be understood literally. All energy is the same as mechanical energy, whether it exists in the form of motion or in that of elasticity, or in any other form. The energy in electromagnetic phenomena is mechanical energy. The only question is, Where does it reside?"[30] The treatise provided a comprehensive account of just where that energy resided, spelling out the mathematical properties of the space-filling ether. It showed, among other things, how waves of electromagnetic energy traveled through the ether at the speed of light, suggesting that light was itself an electromagnetic phenomenon. In Maxwell's hands, the ether became the new focus for the physics of energy.

As far as British physicists were concerned, the defining feature of the ether was that it was a mechanical construct. While Maxwell was clear

[28]J. C. Maxwell to M. Faraday, 9 November 1857. In L. P. Williams, *The Selected Correspondence of Michael Faraday* (Cambridge: Cambridge University Press, 1971), 2: 882.

[29]Quoted in C. Smith, *The Science of Energy*, 227.

[30]Quoted ibid., 232.

The equation of motion of the suspended magnet is

$$\frac{d^2 T}{dt\,d\phi} - \frac{dT}{d\phi} + \frac{dV}{d\phi} = 0, \qquad (10)$$

whence   $A\ddot{\phi} - MG\gamma\cos(\theta - \phi) + MH(\sin\phi + \tau(\phi - a)) = 0.$   (11)

Substituting the value of $\gamma$, and arranging the terms according to the functions of multiples of $\theta$, then we know from observation that

$$\phi = \phi_0 + b e^{-lt}\cos nt + c\cos 2\,(\theta - \beta), \qquad (12)$$

where $\phi_0$ is the mean value of $\phi$, and the second term expresses the free vibrations gradually decaying, and the third the forced vibrations arising from the variation of the deflecting current.

Beginning with the terms in (11) which do not involve $\theta$, and which must collectively vanish, we find approximately

$$\frac{MG\omega}{R^2 + L^2\omega^2}\left\{Hg\,(R\cos\phi_0 + L\omega\sin\phi_0) + GMR\right\}$$
$$= 2MH(\sin\phi_0 + \tau(\phi_0 - a)). \quad (13)$$

Since $L\tan\phi_0$ is generally small compared to $Gg$, the solution of the quadratic (13) gives approximately

$$R = \frac{Gg\omega}{2\tan\phi_0\left(1 + \tau\dfrac{\phi_0 - a}{\sin\phi_0}\right)}\left\{1 + \frac{GM}{gH}\sec\phi_0 - \frac{2L}{Gg}\left(\frac{2L}{Gg} - 1\right)\tan^2\phi_0\right.$$
$$\left. - \left(\frac{2L}{Gg}\right)^2\left(\frac{2L}{Gg} - 1\right)^2\tan^4\phi_0\right\}. \quad (14)$$

If we now employ the leading term in this expression in equations (7), (8), and (11), we shall find t at the value of $n$ in equation (12) is $\sqrt{\dfrac{HM}{A}}\sec\phi_0$. That of $c$, the amplitude of the forced vibrations, is $\frac{1}{4}\dfrac{n^2}{\omega^2}\sin\phi_0$. Hence, when the coil makes many revolutions during one free vibration of the magnet, the amplitude of the forced vibrations of the magnet is very small, and we may neglect the terms in (11) which involve $c$.

766.] The resistance is thus determined in electromagnetic measure in terms of the velocity $\omega$ and the deviation $\phi$. It is not necessary to determine $H$, the horizontal terrestrial magnetic force, provided it remains constant during the experiment.

To determine $\dfrac{M}{H}$ we must make use of the suspended magnet to deflect the magnet of the magnetometer, as described in Art. 454. In this experiment $M$ should be small, so that this correction becomes of secondary importance.

3.4 A page from Maxwell's *Treatise on Electricity and Magnetism* (1873). He is describing William Thomson's method for determining the value of the standard unit of resistance (the ohm) as used by the British Association for the Advancement of Science's Committee on Electrical Standards.

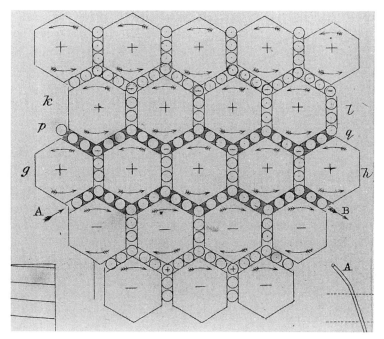

3.5 Maxwell's representation of a possible mechanical structure for the electromagnetic ether.

that his 1862 model of vortices and idle wheels was only a hypothesis and not necessarily an accurate representation of what the ether's structure was really like, there was no question but that the ether had some kind of mechanical structure (figure 3.5). Just as the machines on Britain's factory floors were made up of pistons and pulleys, flywheels and governors, cranks and gears, so was the electromagnetic ether. The conservation of energy was meant to be a theory that applied in the real world. The link between the engines and machines of Britain's factories and the electromagnetic ether was that both could be approached with the same physics, the same theories and models. This was because they had the same kind of structure—they were made of the same kind of things. In this way the high mathematical physics of Maxwell and his successors really did have universal applicability. It explained the factory as much as it explained the ether—and derived its credibility from its capacity to explain both.

In his review of Oliver Lodge's *Modern Views of Electricity* (1889), the French philosopher and physicist Pierre Duhem condemned the British proclivity towards mechanical models: "Here is a book intended to expound the modern theories of electricity and to expound a new theory. In it there are nothing but strings which move around pulleys, which roll around drums, which go through pearl beads, which carry weights; and tubes which pump water while others swell and contract; toothed wheels which are geared to one another and engage hooks. We thought we were entering the tranquil and neatly ordered abode of reason, but we find ourselves in a factory."[31] Lodge, professor of physics at University College Liverpool and an enthusiastic modeler of the ether, would happily have concurred with this assessment. Like his fellow Maxwellian at Trinity College Dublin, George Francis FitzGerald, he took modeling the ether to be a crucial heuristic and pedagogical activity. Since the ether was a mechanical system, trying to construct mechanical models that reproduced its properties was a valuable way of finding out more about it. It was also a useful way of demonstrating its properties to others.

Lodge and FitzGerald were convinced that the Holy Grail of physics would be a purely mechanical model of the ether—one that was not only useful heuristically and pedagogically, but that could be taken to be a true representation of the ether's real structure. According to Lodge, "If a continuous incompressible perfect fluid filling all space can be imagined in such a state of motion that it will do all that ether is known to do; if, simply by reason of its state of motion, it can be proved capable of conveying light and of manifesting all electric and magnetic phenomena which do not depend on the presence of matter; and if the state of motion so imagined can be proved stable and such as can readily exist, the theory of free ether is complete."[32] This was the late nineteenth-century version of the end of physics. Lodge had a likely candidate in mind as well in the form of FitzGerald's vortex sponge model, first devised in 1885. From the 1880s through the end of the century, refining this model into an adequate representation of the real ether was the central goal of British physics. Success would mean the reduction of all physics to mechanics—the science of matter and motion. Maxwell's equations defining the operations of the electromagnetic field could then be rewritten in purely mechanical terms.

By the end of the nineteenth century the physics of energy was triumphant. The doctrine of the conservation of energy was a central and indispensable plank not only in physics but across a whole range of

[31] P. Duhem, *The Aim and Structure of Physical Theory* (Princeton: Princeton University Press, 1954), 70–71.

[32] O. Lodge, *Modern Views of Electricity* (London, 1889), xi.

sciences. This physics of energy was widely accepted as being the physics of the ether as well. The ether in some form or other was generally accepted as the universal medium through which energy traveled (at the speed of light) and in which it was stored. As Pierre Duhem had rather snootily noted, this energy physics was also the physics of the factory. It forged a link in the chain binding the high mathematics of Cambridge analysis to the realities of industrial culture. The conservation of energy was a broad church, however, and not all its adherents understood it in the same way as its high priests. William Robert Grove, in the final edition of the *Correlation of Physical Forces* (1874), was still staking his claim to its discovery. In his obituary of Grove in the scientific magazine *Nature* in 1896, the physicist Andrew Gray, Lord Kelvin's successor to the chair of natural philosophy at Glasgow, acknowledged his theory of correlation as a precursor of the conservation of energy. The phrases "correlation of physical forces" and "conservation of energy" were still interchangeably used in the popular (and even occasionally in the professional) sphere at the end of the century. Energy did nevertheless provide its adherents and practitioners with a powerful tool, not only for understanding nature, but for carving out a niche for themselves as professional scientists in fin de siècle industrial culture.

## Conclusion

The nineteenth century was a period of massive and unprecedented social upheaval. Populations exploded, national borders were redefined, new political systems and ideologies were forged, industrialization and urbanization destroyed old ways of life and created new ones. Natural philosophy's place in this new social order needed to be redefined. This was a highly contested process. Not everyone—not even all men of science—agreed what natural philosophy should look like and what it should say about nature and society. The perspective of, say, an early nineteenth-century German enthusiast for *Naturphilosophie* would in this respect be very different from that of a midcentury advocate of the correlation or conversion of forces, regardless of any superficial similarity in their views about the unity of nature. To realize their competing visions, practitioners of natural philosophy had to forge new spaces and new institutions for themselves. They had to turn their science into something that mattered for their rapidly changing industrial culture. They had to convince their audiences that natural philosophy made a difference. In this process, physics and physicists were transformed. In many ways, the history of nineteenth-century science was one of constant refashioning as natural philosophers and scientists tried to find a secure cultural niche for themselves and their productions.

Physics certainly was transformed by the end of the century. New institutions and laboratories proliferated and chairs were established in new civic universities. It was the dominant science, and its practitioners wielded the power to adjudicate over the proceedings of disciplines far removed from theirs. Physics, thanks to the conservation of energy, was recognized as the foundational discipline. This may seem self-evident to modern sensibilities—what other science should be regarded as the model against which all others have to measure themselves? That position was not purchased without a fight, however. In the early years of the nineteenth-century chemistry might have appeared a better candidate. At midcentury, astronomy still ruled the roost as the exemplar of scientific methodology. Physics in the nineteenth century reorganized itself and reorganized the world around it at the same time. The conservation of energy turned out to be the ideal tool for creating and holding together a new discipline that crossed the boundaries between factories, laboratories, and university studies and lecture halls. Its affiliations with the mathematical world of the mind made lab work respectable for the sons of gentlemen, while its connections with the world of telegraph cables, electric power, and factory engines made it a practical occupation for the sons of trade.

# 4

## The Science of Showmanship

Throughout the nineteenth century, the most visible of the physical sciences in many ways was the burgeoning science of electricity. Electricity provided the technology for a whole range of vivid and spectacular demonstrations of nature's powers, and of man's powers over nature. As the century advanced electricity gave rise to a whole range of new industrial technologies as well. Across Europe and America, many identified their century with unprecedented economic and social progress. Many agreed also that the combined forces of science and industry could be identified as being at the root of this newfound prosperity. For these observers, it was precisely the development of new ways of understanding nature and exploiting her resources that seemed to promise never-ending progress. Electricity in many ways seemed to epitomize this process. New industrial technologies like electroplating provided luxury goods for the growing middle classes; the electric telegraph provided practically instantaneous communication across the globe; by the end of the century electrical energy was providing light and power in households all over the Western world. The electrical future increasingly seemed to promise more. Much of nineteenth-century physics was inextricably embedded in these new industries as well. The science of electricity was very much about making this new power spectacularly visible and making it useful too.

These kinds of links between electricity and the worlds of showmanship and industry were hardly new by the nineteenth century. Since the early eighteenth century spectacular electrical demonstrations of nature's powers had been part of the stock in trade of philosophical showmen. They could use electricity to show off not only God's powers in nature but their own abilities to control those powers. Natural philosophical lecturers, performing in coffeehouses and salons across Europe, vied to produce more and more spectacular experiments. New and more powerful electrical machines were constructed, along with new devices like the Leyden jar to store and concentrate the electric effluvium (as it was often described). Genteel coffeehouse habitués could be amazed, shocked, or even beatified (crowned by a fluorescent halo) by the new electric fluid. As we saw in the introduction, there was an entrepreneurial edge to many of these electrical activities in the eighteenth century as well. Popular lecturers such as John Desaguliers and Benjamin Martin in England and instrument makers such as the Dutchman Martinus van Marum had their eye on attracting potential patrons to make their productions commercially viable. The American Benjamin Franklin's invention of the lightning rod was motivated by practical commercial concerns as much as by philosophical curiosity.

Exhibition played a crucial role throughout nineteenth-century culture for a number of reasons. The rising middle classes flocked to a whole range of public entertainments as the century progressed, from theatrical productions to grand musical soirees or fireworks displays. Following the unprecedented success of the Great Exhibition at the Crystal Palace in 1851, imitations sprang up all over Europe and the United States. By the final decades of the century, international exhibitions and world fairs were annual events in cities throughout the western hemisphere. Other forms of display were also proliferating. As relationships between producers and consumers of goods changed, department stores featuring gaudy displays of commodities on sale became common features of metropolitan streets. Advertising developed new ways of drawing the consumer's attention. Natural philosophers had their place in this world of display. Many earned their living through popular lectures drawing crowds through spectacular experimental demonstrations. They had to compete directly for customers with the theaters, panoramas, dioramas, and magic lantern shows of metropolitan culture. Electricity was a crucial resource for such performances, and electrical experimenters worked hard to find new and spectacular ways of making electricity visible.

The nineteenth century witnessed an upsurge in invention as well. Hopeful entrepreneurs and inventors flocked to the patent offices, aiming to make their fortunes through this or that new improvement or spectacular breakthrough. Arguments raged concerning the relationship between science and the industrial arts. Some, like Charles Babbage, insisted that there was an inseparable dependence. The natural sciences, according to Babbage and his allies, were the powerhouse that generated economic progress through a constant supply of new ideas to be applied in new machines and technologies. Others disagreed, arguing that there was no link between the rarefied ideals of natural philosophy and the grubby business of invention and salesmanship. Electricity had an important role to play in these kinds of debates too. To many, it was the prime example of science applied successfully to the arts. New technologies such as the electric telegraph seemed to demonstrate conclusively the central role of natural philosophical discovery in inventive activity. Again, others disagreed, suggesting that the history of such inventions tended to show they were the products of inspired tinkering by practical men rather than the systematic application of natural laws.

There was, of course, a close connection between these joint concerns with exhibition and utility. There was in the first instance very little practical difference between the process of producing a new piece of electrical apparatus as part of a public lecturer's or demonstrator's box of tricks and inventing a new device with an eye to the marketplace. The devices themselves were as often as not practically identical, or at least closely related. When lecturers put themselves and their demonstrations of nature in action on show, they were frequently quite literally inhabiting the same space as that where hopeful inventors exhibited their own productions. In the Victorian public's mind there was very little difference between a device designed to demonstrate some natural principle and an item of economic utility on sale. Again like the telegraph, the two could often be the same. Crucial cultural events such as the Great Exhibition played an important role in crystallizing such perceived relationships as well. The Crystal Palace was only one of the more impressive spaces where invention and discovery rubbed shoulders.

The place of the physical sciences and of electricity in this culture of exhibition and entrepreneurship was not uncontested or uncontroversial. Many men of science argued determinedly that the sciences had no proper role in such a context. Crass utilitarianism and grubby display were alike beneath the dignity of natural philosophy. Others had no such qualms. Many such arguments in the end boiled down to the simple question of who the man of science—in this case the electrician—was. Was he (he was almost invariably a "he") the disinterested discoverer of natural principles, the polite purveyor of new experimental knowledge to a genteel middle class or aristocratic audience? Or was he the flamboyant showman shocking (sometimes quite literally) his audience with the latest example of man's dominance over nature, demonstrating his mastery over nature through his control of the machines he exhibited? Or was he the hardheaded inventor of revolutionary new technologies, industriously playing his part in transforming the nineteenth-century economic landscape? As the century progressed, the science of electricity was to be a key battleground in resolving such questions.

## Foundations of a New Science

In the early 1780s, Luigi Galvani, the professor of anatomy at the University of Bologna, carried out a series of experiments that demonstrated, he claimed, that there was a specific electricity produced by animal bodies (figure 4.1). He found that when the nerves and muscle of a frog's leg were connected by means of a metallic conductor, the leg twitched. This indicated, according to Galvani, the existence of a flow of electricity running through the dead frog's nervous system. This animal electricity—or galvanism as it was soon to be designated in honor of its discoverer—was to be a source of major controversy. Galvani and his adherents insisted

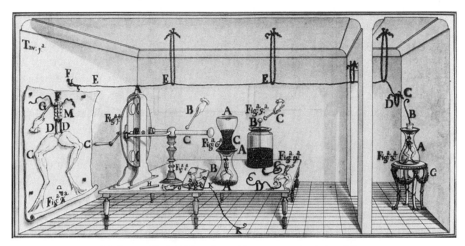

4.1 Some of Luigi Galvani's experiments on animal electricity.

that the source of the electricity was the animal tissue and that Galvani had therefore discovered an entirely novel variety of electric fluid. His opponents, notably Alessandro Volta at the University of Pavia, were just as adamant that the metal in contact with the tissues was the source. The electricity was simply produced by the contact of two dissimilar metals according to this view. All the animal tissue did was facilitate that contact. The dispute raged over two decades as both Galvani and Volta produced experiment after experiment, each proving his own and disproving the other's claims. No one denied the existence of this novel form of electricity. The issue was its origin. Was there, as Galvani claimed, an innate electricity in animal bodies, or was the electricity found in such circumstances merely the result of metals in contact, as Volta asserted?

In 1800, Volta made public a new experiment that he thought was decisive. When a pile of zinc and copper discs was constructed, each copper and zinc pair separated by a paper disc soaked in acid or a saline solution, an electric current flowed from one end of the pile to the other. This was Volta's final model of what happened in Galvani's experiment. No animal tissue was needed, suggesting, of course, that there was no specific animal electricity after all. Volta toured Europe with his voltaic pile, as it soon became called, demonstrating to excited savants at the Institut de France in Paris and elsewhere the power of his new instrument (figure 4.2). Galvani's nephew, Giovanni Aldini, likewise went on a Grand Tour to demonstrate that electricity could be produced from animal tissue without the intervention of metallic conductors. In London he even carried out spectacular electrical experiments on the corpse of an executed felon. The focus of European attention, however, was Volta and his new instrument in its various permutations. It seemed that he had discovered a new and powerfully versatile source of the electric fluid. In 1801, Napoleon, the new French emperor, awarded him a medal for his services to science and to celebrate the grand discovery made on what was, by then, newly conquered French territory. The question of the source of electricity in the voltaic pile remained open, however. Volta and his newfound French allies insisted that it was the contact of the metals. Others, notably among the revolutionary French state's enemies across the English Channel, insisted that it must lie elsewhere.

The rapidly rising star of English science in the 1800s was Humphry Davy, newly arrived at the Royal Institution in London, itself just established to place natural philosophy at the service of the embattled English state by improving agriculture and the industrial arts. Davy seized on the voltaic pile as a powerful new weapon in his armory of chemical

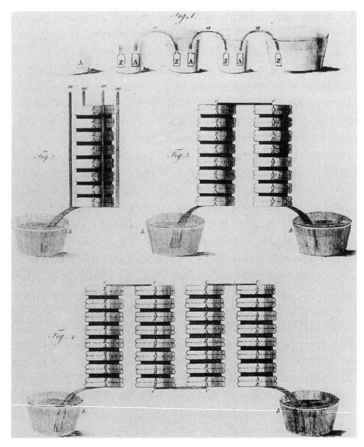

4.2 Some early examples of Alessandro Volta's voltaic piles.

apparatus. Along with other pioneering English experimenters such as William Cruickshank and William Nicholson, he transformed Volta's small-scale device into a giant instrument for chemical demonstration and analysis. Davy used this potent new resource to dazzle and amaze his genteel Royal Institution audience with his capacity to subjugate nature. It provided the foundation for his growing reputation as a consummate philosophical performer. At the same time it provided the evidence for Davy's chemical view of electricity. Davy could use the powerful electrical forces produced by the voltaic battery to tear chemical compounds apart and reveal their constituent elements. Soils could be analyzed, new chemical elements such as chlorine could be isolated by subjecting them to the currents from the Royal Institution's batteries. The implication was

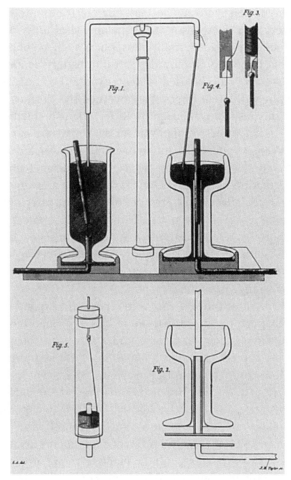

4.3 Michael Faraday's demonstration apparatus showing that a
current-carrying wire could be made to revolve around the
central magnet. Michael Faraday, *Experimental Researches in
Electricity*, vol. 2, plate 4.

that the electric force was chemical in origin as well. Davy claimed that
the voltaic pile's electricity derived from chemical reactions rather than
from the contact of its metals. His French counterparts, particularly those
in the Laplacian camp, still followed Volta in maintaining that metallic
contact was what mattered.

While English, French, and Scottish experimenters continued to debate the respective merits of the chemical and contact theories of galvanic
action, their counterparts elsewhere, particularly in the German lands,

took a different perspective. They saw the electricity produced by the voltaic battery as a microcosm of the metaphysical unity of nature. As we saw earlier, Romantic natural philosophers such as Johann Wilhelm Ritter in Jena hoped to employ the galvanic battery as an instrument of metaphysics. It could be used to demonstrate the fundamental unity of the seemingly different forces that governed the Cosmos. The breakthrough in this respect was made by Ritter's Danish collaborator Hans Christian Oersted, professor of physics at the University of Copenhagen, in 1820. A keen exponent of Kantian metaphysics, Oersted aimed to use the battery to find a link between electricity and magnetism. After careful experimentation he succeeded in showing that a magnetized needle could be made to deflect in the presence of a current-carrying wire. News of the amazing discovery fascinated Europe's philosophical community. Oersted's short Latin publication was rapidly translated into English, French, and German. His experiment was repeated before skeptical audiences (particularly in Paris) as electrical experimenters tried to make sense of this strange new phenomenon.

Some of the most diligent efforts to repeat and expand on Oersted's work were made at the Royal Institution by Davy's laboratory assistant, Michael Faraday. Faraday had come under Davy's patronage following the end of his apprenticeship as a bookbinder. He had joined Davy on his Grand Tour through war-torn Europe in 1813, when his master was invited to Paris to be fêted by Napoleon and awarded a medal for his philosophical discoveries. By 1820, Faraday was starting to experiment in his own right and was anxious to make a name for himself as an independent philosopher. In 1821, in a careful series of experiments he demonstrated that a current carrying wire could be made to rotate around a magnet. His work not only verified Oersted's claims concerning the relationship of electricity and magnetism, it also seemed to confirm that the force from the wire did not act as forces usually did—towards a central point—but that it rotated around the wire instead. Faraday's achievement was not universally acknowledged at the genteel Royal Institution. One of the institution's patrons, William Wollaston, was already engaged in a series of experiments to investigate the apparent rotatory motion of current carrying wires near magnets. It appeared unseemly that a mere laboratory assistant should have preempted his discoveries. The discovery was, however, sufficient to establish Faraday as a philosopher in his own right, a position that he carefully consolidated at the Royal Institution over the next decade.

4.4 Michael Faraday lecturing before an audience including Prince Albert and the prince of Wales at the Royal Institution.

By 1830, Faraday was a fellow of the Royal Society and director of the laboratory at the Royal Institution. He was widely recognized as having stepped into Humphry Davy's shoes as the foremost exponent of natural philosophy to the English upper classes (figure 4.4). In 1831, he embarked on an ambitious experimental program that was to make him one of Europe's premier electrical experimenters as well. In what turned out to be only the first installment of his *Experimental Researches*, presented before the Royal Society on 24 November 1831, Faraday showed that magnets could be used to create electricity. When a bar magnet was inserted into a wire coil and again when it was removed, a brief current was recorded on a galvanometer connected to the coil. Also, when a current was passed through a coil of wire wrapped around a soft iron ring, a current could also be recorded on another, unconnected coil of wire wrapped around the same ring whenever the original current was switched on or off. Faraday called this effect induction, to remind his auditors and readers of the familiar field of ordinary (static) electricity. As in his demonstrations of a decade earlier, Faraday had opened up a whole new field of enquiry into the links between electricity and magnetism, as well as spawning a whole range of new electrical instruments and devices to put the new phenomena on show. Over the next three decades, Faraday produced twenty-nine series of these *Experimental Researches*, translating the results of his endeavors in the Royal Institution's basement laboratory to an increasingly expectant audience.

A few years later Faraday made another breakthrough. In his "third series" he outlined a number of experiments designed to confirm and demonstrate the identity of the electricities. It was still an unresolved issue whether the electricity deriving from a galvanic battery was the same as the electricity derived from an electrical machine or Leyden jar. This was particularly so in that many of their effects seemed very different in scale and kind. Faraday set out to show by careful measurement that the electricities were in fact identical in that the effects of a given electricity could be reproduced with electricities from different sources. The differences usually observed could be attributed to variations between sources in the quantity and intensity of electricity being made available. In further research he established as well the chemical equivalence of electricity. A given amount of electricity used to break down a chemical compound would do so in proportion to the elements contained. When water was decomposed by electricity, for example, twice as much hydrogen as oxygen was given off in keeping with water's chemical composition of two parts hydrogen to one part oxygen. Faraday used this apparently absolute relationship to propose a new way of measuring electricity. He suggested that the amount of gas given off when electricity was passed through a tube of water could be used as an absolute measure of the quantity of electricity. He baptized the new instrument the Volta-electrometer.

By the beginning of the 1840s Faraday was firmly established as one of Britain's foremost (if not the foremost) experimental natural philosophers. He was also beginning to publish some of his private speculations concerning the nature of electricity, force, and matter. He was increasingly convinced that electricity should be regarded as a force occupying the space surrounding conductors rather than as a fluid (or fluids) flowing through the conductors themselves. He elaborated this view in papers such as "Speculation touching Electric Conduction and the Nature of Matter" (1844) and "Thoughts on Ray Vibrations" (1846). His views were bolstered by his magneto-optic experiments of 1845, in which he demonstrated that a ray of polarized light passing through glass along a magnetic line of force would be rotated according to the direction of the line of force. There is an interesting link between the view of matter that Faraday was promoting here—that what mattered was the distribution of lines of force in space—and his pedagogical strategy. Faraday's aim in the lecture theater was to direct his well-bred audience's attention away from the grubby details of the apparatus he used to produce the phenomena. He wanted them to see that nature (not he or his instruments) was doing the work. His view of matter directed attention away from the instruments and towards the space surrounding them as well. His pedagogy and his ontology went hand in hand.

Very few natural philosophers took Faraday's strange views on lines of force in space seriously until they were picked up by James Clerk Maxwell a decade later. Most British electricians maintained the view, implicitly at least, that electricity should be regarded as some kind of imponderable fluid. Their task was to make that fluid visible. In France, a more mathematical approach to electricity developed, drawing largely on the Laplacian tradition of looking at natural processes as resulting from the action of central forces. The initial response in France to Oersted's experiment was to see if it could be fitted into the Laplacian straitjacket. Jean Baptiste Biot and his student Felix Savart reduced the phenomenon to a simple law. Imagining the needle as consisting of molecules of magnetism each with a north and south pole, they wrote early in 1820: "Draw from the pole a perpendicular to the wire; the force on the pole is at right angles to this line and to the wire, and its intensity is proportional to the reciprocal of the distance."[1] Such language had the virtue of preserving the Laplacian insistence on simple forces acting on points (or molecules) in space and transformed the phenomenon into a mathematical generalization.

Biot's fellow Frenchman and adversary André-Marie Ampère, on the other hand, was less wedded to the Laplacian worldview. He used Oersted's experiment to try to break down the distinction between electricity and magnetism. He argued that the best way to understand the way in which electricity and magnetism interacted was to think of the two forces as identical. Magnetism was the result of electricity in motion. Permanent magnets could be regarded as consisting of a number of loops of electric current. The direction of the current in the loop determined the magnet's polarity. Ampère bolstered this view by showing how a current-carrying helix could be made to act like a magnet. He showed that current-carrying wires attracted and repelled each other as well—just like magnets. Ampère—a coconspirator with anti-Laplacians such as Arago, Fourier, and Fresnel—saw himself as the founder of a new science of electrodynamics that moved away from Laplacian shibboleths.

[1] E. Whittaker, *History of the Theories of Aether and Electricity: The Classical Years*, 82.

Carefully contrived experiments were crucial for Ampère's work, both as demonstration and measurement. His claims concerning the electrodynamic nature of magnetism were considerably more credible to his skeptical academical colleagues in Paris once he could show them (as he did on 25 September 1820) that current-carrying helices really could be made to act like feeble magnets.

By the 1830s and 1840s, distinct ways of doing electricity were emerging. Volta's invention of the voltaic pile (precursor of the modern electric battery) provided a powerful new source of the electric fluid. Oersted's experiment of 1820, followed by Faraday's and Ampère's experiments, forged a new, intimate connection between electricity and magnetism. Different languages of electricity were also emerging. Faraday, addressing as he did a primarily lay if socially prestigious audience at the Royal Institution, presented his work in the vernacular. Ampère across the Channel, speaking as he did to his colleagues at the Académie des Sciences, spoke the language of mathematics. Faraday was in any case deeply suspicious of any efforts to express natural philosophy in abstract mathematical terms. This was a common view among British experimenters, who argued that the abstract manipulations of algebraists led the natural philosopher too far away from the phenomena they were meant to be studying. French experimenters tended to take the opposite view, arguing on the contrary that mathematics provided a language of precision that allowed for clear and unambiguous descriptions of real phenomena. What underpinned both languages, however, was an increasing array of experiments and instruments designed to make electricity visible.

### The Technology of Display

While electricity had provided natural philosophers with a source of spectacular demonstrations of nature's powers since the eighteenth century, Volta's invention of the battery and Oersted's demonstrations of the link between electricity and magnetism provided experimenters with the raw materials for a whole range of new technologies. Ways of rendering the electric fluid visible on a grand scale proliferated during the first half of the nineteenth century. There was more to this quest for striking displays of the forces of nature than simply a desire to put on a good show, though as many experimenters were dependent on income earned from lectures for their living this was certainly a consideration. The arrays of batteries, electromagnets, galvanometers, induction coils, magneto-electric machines, and voltameters that made up the technology of display in a very real sense constituted the electrical world as well. They provided models for the operations of natural systems. The cosmos could quite literally be seen as being composed of machines, analogous to the ones that electricians demonstrated at lectures and exhibitions. A conducting sphere rotating around a magnet by means of thermoelectricity, for example, was "obviously analogous to the natural state of the earth"[2] and explained its rotation as the result of the difference in temperature between the equator and the poles. This perception also had an important role to play in sustaining electricians' authority as interpreters of the natural world. By demonstrating their skills in constructing, manipulating, and controlling their instruments, they guaranteed to their audiences their mastery over nature as well.

One of the centerpieces of the technology of display was the electromagnet, invented in 1824 by the English electrician William Sturgeon. Sturgeon,—the author of the analysis of the Earth's rotation mentioned in chapter 3—earned a precarious living through instrument making and lecturing. The electromagnet—a coil of copper wire wound around a soft iron horseshoe—was one of a number of devices he submitted to the Royal Society of Arts, for which he won a prize of thirty guineas and a silver medal. The explicit aim in constructing this portable set of tabletop apparatus was to find ways of making the electric fluid more visible as economically as possible (figure 4.5). To this end, Sturgeon was looking for ways of maximizing the effects produced with his apparatus without a concomitant increase in the size of the source of power. In this process he found that by wrapping a wire coil around a soft iron core, he could dramatically increase its magnetic power. An added advantage of this new device was that the magnetic power could be switched on and off instantaneously simply by connecting or disconnecting the source of electricity. The new instrument could graphically demonstrate the powers of electricity and magnetism and the demonstrator's ingenuity by raising and dropping large weights at will.

Sturgeon's electromagnet was a comparatively small-scale device—literally a piece of tabletop equipment. In other hands, however, the instrument became truly gigantic. The Dutch natural philosopher Gerrit Moll found ways of significantly increasing the power of the electromagnet by rearranging the ways in which the wire was coiled around the arms of the magnet. Innovations introduced by the American experimenter Joseph Henry massively increased the instrument's power. Henry,

---

[2]W. Sturgeon, "On Electro-Magnetism," *Philosophical Magazine*, 1824, **64**: 248.

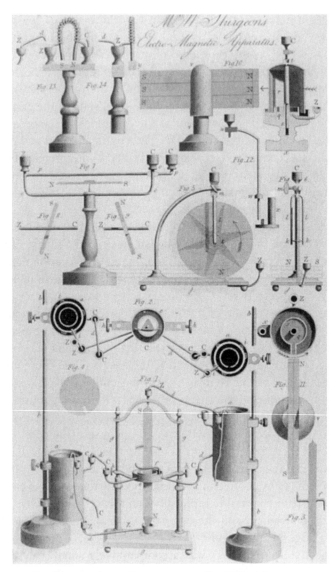

4.5 The electromagnetic table-top apparatus that William Sturgeon
     presented to the Society of Arts in 1824. His electromagnet is shown
     in the top left-hand corner.

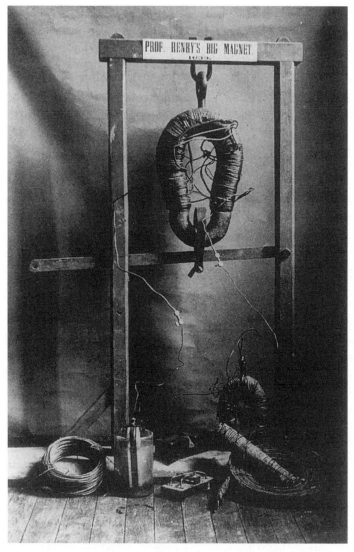

4.6 The massive electromagnet built by Joseph Henry for use in classroom demonstrations at New Jersey College in Princeton.

professor of natural philosophy at the Albany Institute and later at New Jersey College (now Princeton University) found that by packing the coils tightly and varying both the length and thickness of the wire he could significantly augment the magnet's lifting power (figure 4.6). In his first experiments, Henry succeeded in constructing electromagnets that could support between 20 and 40 pounds of weight. Within a few years, however, his skills at electromagnet construction had developed to the extent that he could construct instruments capable of supporting more than 600 pounds using the electricity from a single voltaic pair. The electromagnet he constructed for Yale College in 1831 could sustain 1,600 pounds. Like his English counterpart Sturgeon, and for similar reasons, Henry was concerned to maximize the power of his instruments for the minimum outlay in terms of battery power.

The production of economical and effective batteries was a perennial concern for early nineteenth-century electrical experimenters. Volta's original design of a pile of zinc and copper discs provided only a comparatively feeble current and was soon superseded. Volta himself developed an alternative design—the *couronne des tasses*—in which plates of zinc and copper were placed in cups of acid. Early English battery designers such as William Cruickshank and William Nicholson favored a trough design. A long wooden trough was divided into a number of partitions each containing a plate of zinc and copper. The metal plates were connected and the trough filled with acid to produce a battery of several elements or pairs of metals as required, allowing for the development of a powerful current of electricity. Designers recognized that different kinds of batteries were required for different purposes. An intensity battery consisted of a number of pairs of small plates connected consecutively (in series, in modern terminology). Quantity batteries, on the other hand, consisted of a single large pair of plates. Intensity or quantity batteries were used according to what kind of electrical effects were required for a particular demonstration or experiment. Some of these batteries could be huge. William Pepys at the London Institution, for example, in 1823 had constructed a quantity battery consisting of two plates fifty feet long by two feet wide.

A perennial problem with early battery designs was their constancy. The current in a basic voltaic cell tended to decrease rapidly with time as polarization effects built up. As a result, effective displays of battery power could be carried out only with freshly charged apparatus and could not be sustained for lengthy periods of time. Much effort was devoted to solving this difficulty. In 1836, John Frederic Daniell, professor of chemistry at King's College London, designed the first constant battery. A few years later, in 1839, William Robert Grove, soon to be appointed professor of experimental philosophy at the London Institution, designed a more powerful cell using nitric acid and platinum plates instead of copper. Robert von Bunsen in Germany soon produced his own version of the

Grove cell, using cheap carbon rods instead of the more expensive platinum. This underlines the importance of economy as well as constancy for instrument makers. As John Shillibeer, an English battery designer, remarked, what was needed was a battery that "requires but a little food, and with that little will perform a good honest day's work."[3] More than simply financial imperatives were at stake in this concern with economy (though again, these mattered). Electrical instruments were held to mirror the operations of nature. Since nature was held to operate with due economy, so battery makers aimed at economy in their designs as well.

In order to convince others of the superiority of their battery designs, experimenters needed reliable and generally recognized ways of assessing battery power. One of the earliest instruments designed to this end was the galvanometer, itself a comparatively straightforward application of Oersted's original needle experiment. In a simple galvanometer a magnetized needle was suspended inside a coil of wire connected to a battery. The extent to which the needle deviated was an indication of the battery power. The instrument was first developed by Johann Schweigger, professor of chemistry at the University of Halle, as a way of augmenting the Oersted effect, hence its original designation as an "electromagnetic multiplier." As we saw previously, Faraday in the early 1830s suggested that the amount of gas given off by the decomposing action of an electric current could provide a measure of the quantity of electricity involved. Other experimenters proposed the length of wire that could be rendered red-hot by a current; the length of a spark that could be produced between the terminal wires; or the weight that could be suspended from an electromagnet connected to a battery, as alternative measures of a battery's power. Typically, battery designers lauding their instrument's powers would employ a whole range of different methods of assessment.

A crucial question was what precisely these various methods and instruments should be regarded as measuring. Faraday's assertion that the amount of gas decomposed by the passage of electricity could be taken as an absolute measure of the quantity of electricity passing was subjected to scathing criticism by William Sturgeon. Sturgeon pointed to the range of factors that could in practice affect the decomposition process, such as the area submerged in the decomposing liquid or the distance between the poles. He also denied that there was any such straightforward correlation as Faraday had described between quantity of current and

[3]J. Shillibeer, "Description of a New Arrangement of the Voltaic Battery," *Annals of Electricity*, 1836–37, 1: 225.

decomposition of gases. His main point was that no single method of assessment should be taken as providing an absolute measure. Different methods provided information about different things. Electromagnets or galvanometers provided information about the magnetic powers of a battery, voltameters provided information about the chemical powers, and so forth. In many ways the issue at stake was what was being measured. While Faraday and others wanted "absolute" measurements of electricity, Sturgeon and some of his fellow instrument makers simply wanted ways of comparing the capacities of various kinds of batteries to produce different kinds of effects. Their concern was simply to make electricity visible to greatest effect. This was what mattered for public demonstration.

Rival instrument makers rapidly picked up on the potential of Faraday's experiments on electromagnetic induction as well for the cause of spectacular public exhibition. Efforts were made to replicate his experiments even before their publication in the *Philosophical Transactions* of the Royal Society, much to Faraday's dismay. The Italian experimenters Vincenzo Antinori and Leopoldo Nobili published their own experiments on induction before the end of 1831, directing their efforts in particular to broadening the range of visible electrical effects that could be produced with the induced current. Faraday himself soon designed an apparatus that allowed him to demonstrate the production of an electric spark from the induced current to his Royal Institution lecture audiences. In Paris, the prominent instrument maker Hippolyte Pixii set about producing an instrument that could produce an extended current rather than the short-lived bursts that Faraday had detected. Pixii's machine, in which a horseshoe electromagnet was rotated in front of the poles of a horseshoe magnet, could be used to decompose water into its constituent gases, for example—an effect that required a lasting current. Faraday's transient effect was in the process of being transformed into something robust, reliable, and replicable.

A similar effort to build a machine for generating a continuous current by means of electromagnetic induction was produced in 1832 by the American instrument maker Joseph Saxton, then working in London at the National Gallery of Practical Science, Blending Instruction with Amusement, or the Adelaide Gallery as it was popularly and understandably abbreviated. Saxton was soon embroiled with Pixii in a priority dispute concerning their respective inventions, which was only resolved by a public contest at the Adelaide Gallery in which both the Pixii and the Saxton machines were put through their paces. It was not long before yet another such machine, developed by the London instrument maker

Edward Clarke, was entered into contention. By the mid-1830s, machines such as these were sufficiently reliable to produce the whole range of effects from an induced current. Shocks, sparks, chemical decomposition, electromagnetism could all be produced at the turn of a handle. Instrument makers learned that, just as with electromagnets, the length and thickness of wire in the coil could be varied to produce different effects to best advantage. Short, thick wires produced quantity effects, while long, thin wires were best for intensity. The priority disputes that surrounded each announcement of a new effect produced by means of the induced current underlines the importance of the technology of display to electricians' culture. It mattered a great deal who could legitimately claim the property rights to such productions.

Another electromagnetic apparatus developed during the mid-1830s to exploit the potential for display of induced currents was the induction coil. Consisting of two coils of wire, one placed inside the other, and both wound around a central iron core with one coil connected via some switching mechanism to a battery, the induction coil could be used to magnify the electrical effects that could be produced from a comparatively small battery. The first such devices were produced by the Irish priest and natural philosopher, Nicholas Callan, at the Catholic seminary of St. Patrick's College, in Maynooth near Dublin. Induction coils in particular had the advantage of being comparatively small and easily transportable. By the 1840s they were increasingly popular as means of administering electricity for medical purposes. Such devices were commonly sold with a clockwork ratchet or electromagnet attachment to accomplish the automatic switching on and off of the battery current without the need for constant manual intervention to ensure a continuous flow of electricity. From the 1850s onwards, more powerful and larger induction coils, commonly called Rühmkorff coils after their inventor, the German Heinrich Rühmkorff, were increasingly employed for the production of large currents of electricity. They were particularly useful for the production of large electric sparks for spectroscopic analysis and for the study and display of electrical discharges through vacuum and low-density gases.

By the 1840s and 1850s instruments of all kinds to display and show off electrical effects were ubiquitous. Devices like Barlow's wheel or Marsh's pendulum, along with a whole range of other electrical paraphernalia, were to be found in any philosophical instrument maker's catalogue in the United States, Britain, France, and the German states. Barlow's wheel, invented by Peter Barlow, the English mathematician and professor at the Woolwich Royal Military Academy, demonstrated electromagnetic rotation by means of a copper disc suspended between the poles of a magnet. When a current was passed along the radius of the disc it rotated. Marsh's pendulum, invented by James Marsh, a Woolwich instrument maker, worked on a similar principle. In Ampère's cylinder, an entire voltaic cell, cunningly mounted around a central magnet, could be made to rotate on its own axis. These devices were not meant exclusively for laboratory or even lecture theater use. They were designed for a wider public consumption as well. Parlor game tricks involving electricity had a long pedigree by the 1840s and 1850s. A favorite was the Venus kiss, where a girl—suitably electrified and sitting on an insulated stool—challenged her male admirers to kiss her. The result, of course, was shocking. These philosophical toys, as they were commonly known, were additions to a repertoire of electrical showmanship stretching well back into the eighteenth century.

Despite the apparent frivolity or ephemerality of some of the electrical technology of display, its importance in understanding the culture of electrical science at midcentury is clear. For many if not most electricians, the business of designing and demonstrating instruments that could be used to make electricity visible was constitutive of the science of electricity. Quite simply, as practicing electricians this is what they spent their time doing. By producing such instruments they were quite literally reproducing nature. It is certainly the case that it was through devices and instruments such as these that the mass of the public both in America and Europe encountered and made sense of the science of electricity. As the case of the induction coil and its descendant the Rühmkorff coil illustrates very well, the extensive electrical technology of display developed by midcentury also constituted the direct ancestors—often very little changed—of later nineteenth-century laboratory teaching apparatus. In many ways the age of classical physics rested on a base provided by early nineteenth-century electricians' technology of display.

## Electricity for Sale

As noted at the beginning of this chapter, there was an intimate connection between electrical exhibition and entrepreneurship. The technology of display could be and was adapted to put electricity to work in a very real sense. An important part of the rationale for the emphasis on making various machines designed to make electricity visible was that the result was to demonstrate the economy of nature as well. Hopeful inventors quickly picked up on the possibilities of putting nature's economy to work for their personal benefit also. Regardless of strictures from high-minded gentlemen of science who believed that any effort to turn science into profit was beneath their dignity and the high standing of their vocation, inventors flocked to the patent offices with a plethora of schemes to turn electrical gadgetry into hard cash. The batteries, coils, electromagnets, and measuring instruments that constituted the stock in trade of the electrician could be exploited to make electrical science a serious commercial proposition. By midcentury, electricity was being used to produce cheap luxury goods for the middle classes, to communicate practically instantaneously over massive distances, to power locomotives, and to illuminate city streets.

Electrometallurgy was the first successful commercial technology (or family of technologies) to be developed from the electricians' technology of display. In its simplest form, this was a process whereby electricity was used to coat an artifact made from some electrically conductive substance with a layer of more expensive or more durable metal, usually silver. The technique was, in many ways, a by-product of efforts during the 1830s to improve the performance of electrical batteries. In a Daniell cell, in which the copper plate of the battery was submerged in a solution of copper sulfate, it was noticed that the copper reduced from the sulfate solution while the battery was active tended to coat the copper plate and that in some circumstances it could be peeled off to produce a relief copy of the plate to which it had adhered. The refinement of this process led to two electrometallurgical technologies: electroplating and electrotyping. In electroplating, a layer of more expensive metal could be coated onto a less expensive metal, providing a way of mass producing luxury goods such as silverware for the middle classes. In electrotyping, the layer of metal building up on the plate would be removed to produce a relief image. This provided a cheap way of mass reproducing images for printing, for example.

Several individuals laid claim to being the inventors of electrometallurgical processes. Moritz Hermann von Jacobi in St. Petersburg claimed priority in invention of the process he named "Galvanoplastik." A Liverpool entrepreneur, Thomas Spencer, claimed for himself the honor of having been the inventor of electrotyping. The disputes surrounding the question of priority, in England in any case, were particularly virulent. Spencer was accused of having stolen other people's work. Some went so far as to suggest that there was literally nothing to have invented, since the whole technology was simply a spin-off from a well-established and recognized side-effect of the action of constant electrical batteries. The passions involved in these debates serve to underline, however, both the importance and the sensitivity of such issues for contemporary electricians' culture. Being able to claim priority in invention of a process such as this could, in some circles at least, provide as much kudos as a claim to philosophical discovery. This was a feature of practical electricians' concerns with the minutiae of technical processes. The first "strictly electro-metallurgical patent" (as it was described by Alfred Smee in his history of the process) was granted however, to a Birmingham merchant, James Shore, in March 1840. Shortly afterwards the cousins George and Henry Elkington were granted a patent for various electroplating processes. Within a decade, electrometallurgy was big business.

The basic technologies of electric telegraphy also had their origins in the machines and instruments making up the technology of display. The early English telegraph pioneer William Fothergill Cooke was inspired to consider the possibilities of using electricity to transmit signals across large distances in 1836, when he attended a lecture at Heidelberg, where a device for displaying electrical effects over long wires, invented by the Russian diplomat Pawel Schilling, was demonstrated. Like many others, however, Cooke soon found that there was a big difference between getting such a device to work on the small scale of a laboratory or lecture theater and making it work in the outside world. He was soon collaborating with the experimental philosopher Charles Wheatstone, professor of natural philosophy at King's College London, in an effort to turn his demonstration devices into a robust and practical technology. Wheatstone had himself been working on the possibilities of exploiting electricity for long-distance communication and had been working on the problem of making electrical effects visible at a distance. In particular, he was aware of Joseph Henry's work on increasing the power of electromagnets—a crucial part of telegraph technology. Henry had visited London only recently on a European tour to acquire philosophical instruments for New Jersey College and had demonstrated his experiments to Wheatstone in person. He knew, therefore, of the importance of winding the coils on an electromagnet properly to maximize their power.

The key to Wheatstone's success in making telegraphy work over distances, however, was his knowledge of the German experimenter and mathematician Georg Simon Ohm's experiments. In 1827, Ohm, a schoolteacher at the Jesuit Gymnasium at Köln, had published a number of experiments on the relationship between current strength and exciting force in current-carrying wires. Ohm's law established that current strength was equal to the exciting force divided by a constant he designated "resistance." Ohm's work was not widely read at the time. It was not published in English until 1841, for example. Wheatstone could read German, however, and recognized that Ohm's work was what he needed in order to understand the problems of transmitting electricity over long distances, as required to build a successful telegraph. It was through telegraphy that the new electrical terminology of currents, potential differences, and resistances came to replace the older terminology of quantities and intensities, though many electricians, like Michael Faraday, strenuously resisted the newfangled concepts. When Wheatstone became embroiled in a series of disputes with his erstwhile collaborator Cooke over their respective roles in inventing the electrical telegraph, it was to his understanding of Ohm's work that he pointed to demonstrate the privileged knowledge that made the telegraph possible. In the meantime, however, Wheatstone and Cooke acquired an English patent for their electric telegraph in June 1837.

In the United States, Samuel F. B. Morse was also working on the possibility of transmitting messages over long distances by means of electricity. Like Cooke, he had stumbled on the idea after encountering examples of electrical demonstration devices during a trip to Europe. In his own famous words: "If the presence of electricity can be made visible in any part of the circuit, I see no reason why intelligence may not be transmitted instantaneously by electricity."[4] In many ways, this notion of exhibition at a distance was exactly what telegraphy was about and underlines its dependence on the technology of display. Like Wheatstone and Cooke, Morse too found himself in difficulties over the problem of how to get electrical effects to work through long wires. Again, Joseph Henry's work on electromagnets proved to be the key to making this version of the telegraph work. The basic principle of the telegraph was quite straightforward. All that was required was a source of electricity (like a battery), a circuit breaker of some kind to enable specific signals to be sent, and a way of making the signal visible. Making such apparatus work, however, required skill, ingenuity, and a detailed knowledge of the ways electrical instruments worked in practice as well as entrepreneurial acumen. In 1840, Morse and his backers were awarded a grant of $30,000 by the U.S. Congress to build a test line between Baltimore and Washington, D.C. Morse's success in America, along with Wheatstone's and Cooke's in Britain, made the telegraph a reality.

[4]E. L. Morse, *Samuel F. B. Morse: His Letters and Journals* (New York, 1914), 2: 6.

A few years after Morse's deployment of an electric telegraph on the Washington-to-Baltimore line, the U.S. Congress awarded a substantial grant ($50,000 in this case) to another electrical project. Charles Grafton Page, a Harvard graduate and official at the U.S. Patent Office, had put forward to Congress a proposal to build and test an electric locomotive. Like the telegraph, early electromagnetic engines had their roots firmly in the technology of display. The first motors had their origins in William Sturgeon's electromagnet. The electromagnet's capacity to rapidly switch its magnetic power on and off raised the possibility of producing a motive force by means of electricity. A number of electricians and instrument makers constructed different kinds of electromagnetic motors during the 1830s. Joseph Henry in the United States developed two kinds of motors—rotatory and reciprocating—for use in classroom demonstrations. Francis Watkins, a London instrument maker, had electromagnetic motors of his own design on sale from the mid-1830s onwards. William Sturgeon also developed his own version of the engine, as did William Ritchie, professor of natural philosophy at the Royal Institution.

A good example of the ways in which the possibilities of electrical locomotion could enthuse inventors is the case of Thomas Davenport, a Vermont blacksmith. By his own account, Davenport was completely untaught in electricity until he came across one of Joseph Henry's electromagnets during a visit to some iron works in Crown Point, New York. Excited by the possibilities of electromagnets for producing motive power, he set about learning all he could about the science and practice of electricity and was soon constructing massive electromagnets of his own. According to a possibly apocryphal account, being short of funds he used silk from his wife's wedding gown as insulation for the wires (Henry was also reported to have used his wife's silk underwear as a source of insulating material for his early electromagnets). By the mid-1830s he was touring the eastern seaboard of the United States exhibiting a rotatory electromagnetic engine and attempting to acquire funds to purchase a patent, which he eventually did in 1837. Settling in New York, he financed his inventive activities by exhibiting his engine to presumably less than enthusiastic crowds before returning penniless to Vermont in 1842. In Russia in the meantime, an electromagnetic engine was famously put on show by Jacobi, who succeeded in propelling a boat along the River Neva by means of its powers.

Inventors and pundits alike were optimistic that transforming relatively small engines like those of Davenport or Jacobi into something capable of performing useful work on a truly commercial level was simply a matter of scale. Making an economically viable engine was simply a matter of bigger electromagnets. Commentators were hopeful that "half a barrel of blue vitriol, and a hogshead or two of water, would send a ship from New York to Liverpool."[5] A systematic effort to investigate this potential was soon carried out by James Prescott Joule by means of detailed experiments aimed at assessing the "duty" of electromagnetic engines—"duty" being an engineering term for the amount of work done for a given quantity of fuel consumed. Joule soon came to the pessimistic conclusion that electromagnetic engines could never outperform those powered by steam and expanded his researches to embrace the study of engine efficiency more generally. Joule's pessimism had little immediate effect, however. In 1842 Robert Davidson, an instrument maker from Glasgow, was carrying out experiments financed by the Edinburgh and Glasgow Railway Company on electric locomotion. Charles Grafton Page in the United States was using his congressional funding to good effect to build an electrical locomotive.

While most electrical patents entered during the 1840s concerned electrometallurgy and telegraphy, an increasing number detailed improvements in ways of providing illumination. Edward Staite in England applied for a number of patents covering his electric arc light, which used the spark between two points of charcoal as the source of light. Like many such entrepreneurs, Staite was a consummate publicist of his new inventions. In 1848 he held a grand exhibition in Trafalgar Square in London, in which his latest light was put into action so impressively that "the Nelson column, which was selected as the principal point, [was] frequently as conspicuous as noonday."[6] In 1849, there was even a new ballet, *Electra*, performed in London and specifically commissioned to show off the brilliance of Staite's electric arc light (figure 4.7). From the 1850s onwards, arc lights like the ones developed by Staite were an increasingly common feature of theatrical performances. The new lights were striking in their verisimilitude. When the French engineers Lacassagne and Thiers put their arc lighting system on show in Lyon, pundits marveled at the light, which was "so strong that ladies opened up their umbrellas—not as a tribute to the inventors, but in order to protect themselves from the rays of this mysterious new sun."[7] Requiring high-intensity currents as they did, arc lights also provided a commercial use for induction coils that

[5] *Mechanic's Magazine*, 1837, 27: 405.

[6] *Patent Journal*, 1849, 6: 80.

[7] Quoted in W. Schivelbusch, *Disenchanted Night*, 55.

4.7 The ballet *Electra* at Her Majesty's Theatre in London in 1849. With electric lights.

could be used to give high intensity from comparatively small electric batteries.

Enthusiasm concerning the economic possibilities of electricity was rife from the 1840s onwards. The success of telegraphy in particular seemed to augur well for future developments. Alfred Smee enthused that "to cross the seas, to traverse the roads, and to work machinery by galvanism, or rather electro-magnetism, will certainly, if executed be the most noble achievement ever performed by man."[8] William Robert Grove, in his inaugural lecture at the London Institution, itself established to promote the alliance of science and commerce, similarly hailed the power of electricity: "Had it been prophesised at the close of the last century that, by the aid of an invisible, intangible, imponderable agent, man would in the space of forty years, be able to resolve into their elements the most refractory compounds, to fuse the most intractable metals, to propel the vessel or the carriage, to imitate without manual labour the most costly fabrics, and, in the communication of ideas, almost to annihilate time and space;—the prophet, Cassandra-like, would have been

[8]Quoted in I. R. Morus, *Frankenstein's Children*, 184.

laughed to scorn."[9] Even before midcentury, electricity's past seemed to bode very well indeed for future triumphs of man's powers over nature. It seemed to be only a matter of time before electricity not only provided the key to unlocking nature's secrets, but established itself as the ultimate source of continuing economic power and progress as well.

## Science on Show

Electricity's technology of display and the culture of entrepreneurship literally shared the same cultural space in the Victorian era's exhibition halls and galleries. From the 1830s onwards, a number of galleries of practical science appeared in London and in some provincial cities. These were places where the Victorian public could go to see the latest developments in science and the arts, displayed for their edification. Shows and exhibitions of various kinds were staples of Victorian popular culture. The metropolitan public could sample a whole range of enlightening entertainment. Magic lantern shows provided glimpses of natural and manmade curiosities of various kinds. Dioramas and panoramas transported the paying customer to exotic and historic times and places. One of the specialties of London's Regent's Park Colosseum, for example, was a huge panorama of the city, viewed as it would be seen from the dome of St. Paul's cathedral. A range of exhibition halls catered for a wide variety of tastes and interests. Exhibitions of scientific and technological artifacts and processes took their place in this context. They were aimed at the kind of clientele that attended other forms of exhibition. In cities such as London, Paris, and Philadelphia, natural philosophical entertainments were part of metropolitan life. Electricity played a key role in many of these exhibitions. As the century progressed and national and international exhibitions proliferated, electricity continued to be crucial. Exhibitions were crucial for electricity as well. They were where, for most of the century, the public went to see and admire electricity in action.

Electrical entertainments came in all kinds of guises. They could be quite formal and elite occasions such as the Friday evening discourses at the Royal Institution in London, presided over by Michael Faraday. At these affairs, prominent men of science would be invited to demonstrate the latest discovery, invention, or curiosity to an audience largely composed of the cream of London society and the metropolis's scientific elite. Faraday himself was a frequent and popular performer, demonstrating

[9]W. R. Grove, *On the Progress of the Physical Sciences* (London, 1842), 24.

the latest of his electrical discoveries. Less formal, but almost as prestigious, were the occasional gatherings organized by John Peter Gassiot, a wealthy wine merchant, enthusiastic electrician, and treasurer of the London Electrical Society. When notable foreign natural philosophers visited, such as Auguste de la Rive in 1843, Gassiot hosted "electrical soirees," where the latest and most spectacular electrical experiments were on show. Such events were in some ways extensions of the long-standing tradition of performing crucial experiments before prominent witnesses so that their authoritative presence could underwrite the experiment's credibility. Events such as these, however, were at the higher end of the social spectrum. Most of the public witnessed electricity in less exalted company.

The National Gallery of Practical Science, Blending Instruction with Amusement, known simply as the Adelaide Gallery, was established between Adelaide Street and Lowther Arcade on the Strand in London in 1832. Its founder, Jacob Perkins, was an American inventor and entrepreneur who had settled in London a decade or so previously. A native of Philadelphia, Perkins was familiar with Peale's Museum of Natural Science and Art, founded by Charles Willson Peale as a repository for natural historical and philosophical curiosities of all kinds. Perkins may well have had Peale's Museum in mind when he set about founding his own gallery, initially designed to showcase his own inventions but soon expanded to encompass the arts and sciences generally. Electricity was an important feature of the gallery's exhibitions. Perkins had hired Joseph Saxton, another recent Philadelphian arrival in London, as the gallery's instrument maker. Saxton's time was largely devoted to electrical matters, such as the magneto-electric apparatus discussed earlier in this chapter. The gallery as a whole was famous as a place where there "were artful snares laid for giving galvanic shocks to the unwary,"[10] including, according to one possibly apocryphal tale, the duke of Wellington.

The Adelaide Gallery soon had a competitor in the Royal Polytechnic Institute, which opened its doors on Regent Street in 1836. Similarly designed to attract the paying public through exhibitions of the latest in invention, one of the polytechnic's star attractions from the early 1840s onwards was a custom-built Armstrong hydro-electric machine. These devices, invented by the industrialist W. G. Armstrong, exploited the capacity of steam released from a high-pressure boiler to produce static

[10] Quoted in W. H. Armytage, *A Social History of Engineering* (London: Faber & Faber, 1961), 146.

electricity. The polytechnic's machine could produce electric sparks a spectacular twenty-two inches in length. By the 1850s, London had another commercial exhibition hall for the arts and sciences in the Royal Panopticon of Arts and Sciences on Leicester Square. Its proprietor was Edward Clarke, previously a philosophical instrument maker and himself a prolific inventor of magneto-electric gadgetry during the 1830s. In the provinces, the Royal Victoria Gallery for the Encouragement and Illustration of Practical Science (usually called simply the Royal Victoria Gallery) was established in Manchester in the late 1830s. William Sturgeon was hired as superintendent and experimented there with, among others, the young James Prescott Joule. As the example of the inventor Thomas Davenport suggests, such exhibitions were increasingly common in the United States as well.

Exhibitions such as these in which the paying public (the usual fee in London was one shilling) came to ponder natural philosophical curiosities in the same space in which they could witness the latest invention or industrial product had a very important effect on the way in which sciences such as electricity and its products were made sense of. To a very large degree, these were the places where the broader public encountered electricity as well as other scientific artifacts and devices. The context in which they saw science in action, therefore, was one in which commodities were on show. The machinery on show at exhibition halls such as the Adelaide Gallery or the Royal Polytechnic Institute were commodities to be bought and sold. They were not for sale at the exhibitions, but the purpose for which their inventors or owners had put them on show there was straightforwardly commercial. They were there to be advertised, to attract potential buyers and customers. This, therefore, was the context for electricity at the exhibitions as well. This was clearly the case for the electric telegraphs, the products of electrometallurgy, and the prototype electromagnetic engines that went on show from the 1840s onwards. It mattered for other, less avowedly commercial, electrical productions as well. Electricity at the exhibition was being turned into a commodity itself, just like the objects surrounding it.

Nineteenth-century exhibition culture in many ways reached its zenith with the Great Exhibition of the Works of Industry of all Nations, held at the Crystal Palace (especially designed for the occasion by Joseph Paxton) in London's Hyde Park in 1851. The Great Exhibition had its precursors, notably in France, where a series of national exhibitions took place regularly in Paris between 1798 and 1849. The original impetus for organizing the Great Exhibition came from the Royal Society of Arts, which had itself been organizing small exhibitions of arts and industry for several years. Under the patronage of Prince Albert, Queen Victoria's husband, the aim was to exhibit on a grand scale the industrial progress of mankind. The exhibition was going to provide a visual instantiation of the grand principle of the division of labor and provide an impetus to international competition and commerce. The exhibition was also designed to instantiate the relationship between science, industry, and art. As Prince Albert put it, "Science discovers these laws of power, motion, and transformation; industry applies them to the raw matter which the earth yields us in abundance, but which becomes valuable only by knowledge. Art teaches us the immutable laws of beauty and symmetry, and gives to our productions forms in accordance with them."[11]

Electricity was well represented at the Great Exhibition. Electric telegraphs of various kinds were among the more common electrical exhibits. Albert himself had drawn attention to the way in which in the new progressive era, "thought is communicated with the rapidity, and even the power, of lightning."[12] Queen Victoria was also duly impressed by the powers of the telegraph, noting in her diary after a visit to the Crystal Palace that "it is the most wonderful thing . . . Messages were sent out to Manchester, Edinburgh &c., and answers received in a few seconds—truly marvellous!"[13] Also on show were a spectacular array of examples of the electroplater's art, mainly supplied by Elkingtons. Various examples of electromagnetic motors were also on show, notably a new design by the Danish inventor Soren Hjorth, who was awarded a prize for his exhibit. There was an electromagnet constructed by James Prescott Joule on show as well, capable of supporting a weight of more than a ton. Edward Staite had examples of his electric arc lights on show. One of the more visible and spectacular electrical exhibits was Charles Shepherd's electric clock, which was prominently displayed in the Great Transept of the Crystal Palace. The main clock was 1.5 meters in diameter and kept time in synchronism with two others placed elsewhere in the building, all powered by a battery of Smee voltaic cells. Voltaic batteries of various kinds were also on display.

The Great Exhibition's success inaugurated a new era of increasingly spectacular international exhibitions throughout the second half of the nineteenth century. Cities and nation-states vied to provide the most

[11] Quoted in R. Brain, *Going to the Fair*, 24.

[12] Quoted ibid., 24.

[13] Quoted in K. Beauchamp, *Exhibiting Electricity*, 84.

successful performance. The scale that such exhibitions aimed at is illustrated by the Paris Universal Exposition of 1867, whose site at the Champ de Mars covered forty-one acres. The exhibition's main building, the Palais du Champ de Mars, was designed by Gustave Eiffel, who later in the century was to design the Eiffel Tower as part of another Parisian international exhibition. Cyrus Field was awarded a grand prize for his work on the recently completed transatlantic telegraph cable, parts of which were on show. The Vienna International Exhibition of 1873 staged a massive public demonstration of the motive power of electricity. The Palace of Industry featured machines by the Gramme company, generating electricity, powering machine tools, and lifting water. At the Berlin International Exhibition of 1879 the exhibits included a Siemens & Halske electric traction locomotive that could carry eighteen passengers around 300 meters of narrow-gauge circular track. More than 100,000 passengers took the trip around the track during the exhibition.

Electricity was particularly visible in the increasing number of American exhibitions held during the last quarter of the nineteenth century. The Philadelphia Centennial Exhibition of 1876 featured a number of telegraphs on display as well as a repeat by the Gramme company of its Vienna display. The highlight, however, was the first display of Alexander Graham Bell's telephone, which won a prize at the exhibition. By the time of the World Columbian Exposition in Chicago in 1893 truly spectacular electrical displays were increasingly a staple of such events. The Electricity Building was lit by 120,000 electric lights (figure 4.8). Visitors could travel from building to building around the site by electric railway. Edison and the Westinghouse Company battled fiercely for the privilege of providing the power plant to drive the exhibition's electrical exhibits. At the California Midwinter International Exposition in San Fransisco's Golden Gate Park a year later, a copy of the Eiffel Tower, which was built for the French exposition of 1889, was constructed. Unlike the original however, this tower featured some 3,200 colored electric lights as well as a powerful searchlight mounted on top. Being seen at the exhibitions was becoming increasingly vital for budding electrical entrepreneurs and inventors. These were the places where electricity encountered its publics.

By the end of the nineteenth century, exhibitions like these were, therefore, crucial forums for electricians and their publics alike. The fight between Edison and Westinghouse (which Westinghouse won) for the honor of electrifying the Columbian Exposition is a good illustration of the extent to which fin de siècle electrical concerns valued the opportunity such events afforded them of putting their wares before the public.

**4.8** The spectacular central display in the Electrical Building at the World
Columbian Exposition in Chicago in 1893.

Exhibitions, however, were important for electricity and electricians for
reasons beyond the opportunity they afforded inventors and public to
display and admire electrical commodities. By providing a showcase for
electricity they provided a showcase for electricians as well. Prominent
men of science such as Lord Kelvin and Hermann von Helmholtz acted as

jurors on such occasions, highlighting their role as arbiters of progress for late Victorian industrial culture. Electrical experimenters such as James Clerk Maxwell used the exhibitions to survey the latest available equipment for their laboratories. International exhibitions were also occasions for international congresses of scientists. At the Columbian Exposition in 1893, the International Electrical Congress took place as well. They completed the work begun at the previous congress in Paris in 1881 (itself also associated with an electrical exhibition) of establishing secure standards for electrical measurements.

Late nineteenth-century inventor-entrepreneurs often represented themselves as flamboyant characters. In many ways showmanship seems to have been part and parcel of the business of electrical invention. A good example is Nikola Tesla, the Serbian immigrant to the United States who made a particular name for himself as an inventor and showman. Tesla's public lectures were a byword for dramatic display. His high-potential, high-frequency electrical apparatus could produce a whole array of spectacular lights and amazing sparks and effects of all kinds (figure 4.9). The highlight of Tesla's performances was when he placed himself in the circuit of his electricity-generating equipment, holding illuminated

4.9 Nikola Tesla showing off with one of his gargantuan high-frequency, high-potential induction coils.

lightbulbs in his hands and passing sparks between his fingers. Literally making himself a part of his invention was ideally calculated to demonstrate his own mastery over it. The French physicist Arsène d'Arsonval, like Tesla known for his researches into electrical effects at high potentials and frequencies, gave demonstrations in which he made himself part of his experimental apparatus as well. Other electrical inventors such as Thomas Edison in the United States and Sebastian di Ferranti in England similarly fashioned themselves through exhibition. Edison was certainly very conscious of the role his image as the "wizard of Menlo Park" played in bolstering his status as inventor. In a way, electrical inventors were putting themselves as well as their inventions on show.

From the galleries of practical science of the early Victorian years through to the massive and flamboyant international exhibitions of the nineteenth century's closing years, exhibitions were crucial for the science of electricity and for electricians themselves. These were preeminently the places were electricians (and a whole range of other men of science) placed themselves and their productions before the public. Exhibition throughout the century had a central role to play in defining electricity's place in culture. Not only did the electrician William Sturgeon lecture at the Adelaide Gallery and later the Royal Victoria Gallery in Manchester, but his instruments and inventions were on show there as well. The same could be said of Edison's, Tesla's, and even Lord Kelvin's appearances at international exhibitions in the 1880s and 1890s. Electricity as a science and as a string of ever more spectacular inventions was made sense of by the public—placed in context—in terms of the places where it appeared. In the nineteenth century that place was the exhibition. The *Telegraphic Journal and Electrical Review* editorialized in 1892 that "[i]t would be interesting if we could know how the future historian will deal with an institution which is peculiar to the nineteenth century. Commencing with the second half of the century, we have had International, General and Special Exhibitions of all kinds. Bazaars and marts are old enough, but an exhibition, though allied to both, is neither one nor the other, and no preceding institution will be found to exactly compare with it."[14]

## Conclusion

The science of electricity underwent a massive transformation during the first half of the nineteenth century. As new ways of producing electricity

[14] *Telegraphic Journal*, 1892, 30: 120.

proliferated, there were more and more places where electricity and its products could be encountered. Novel electrical technologies, experiments, and instruments made up a new world to be explored and articulated. New sources of electricity raised questions about the identity of electricity, for example. Was the electricity generated by a voltaic battery or an electromagnetic machine the same as that derived from a static electricity generator? In particular, electricity provided new terrain for experimenters anxious to make their reputations. Humphry Davy and Michael Faraday in England, Hans Christian Oersted in Denmark, and André-Marie Ampère in France, to cite only a few examples, forged careers and names for themselves as natural philosophers by means of electrical experiments. Thus, they were instrumental in forging meanings for electricity as well. The new science produced through electricity was contested territory. Electricity was a fluid, it was a force; it was evidence for the unity of nature, it was just one more imponderable power; it was the product of practical experiment, it was the product of abstract mathematical reasoning. Whatever electricity was, all the nineteenth-century protagonists agreed that it was well worth fighting for.

Nineteenth-century commentators were certainly aware of the central role exhibitions played in the century's public life. The *Telegraphic Journal* concluded its discussion of exhibitions with the observation that "the institution existed in the latter half of the nineteenth century, because it was one suited to the requirements of the period."[15] Exhibitions provided a way of bringing science, scientists, and scientific productions inside public culture. In many ways they were expressions of late Victorians' confidence in their capacity to transform nature and culture through technology. Electricity was key in these temples to progress. In many ways it was the absence of a good account of what electricity really was that made it so attractive. In a joke making the rounds at the time, a college professor asked a student what electricity was: "The student hesitated, and tried to think of an answer, but in vain, it was no use. He could not recall it, but in self-defence said, 'I did know, but have forgotten.' The professor replied, 'This is terrible. The only man who knew what electricity was has forgotten!'"[16] It was this mysterious quality that made electricity so amenable as a conduit for progress. Its effects could be put on show in spectacular fashion despite the uncertainty surrounding their

---

[15] Ibid.

[16] *Telegraphic Journal*, 1886, **18**: 281.

origins. Seeing electricity at work made tangible the prospects of future power when its secrets finally were revealed.

In many ways exhibitionism became more culturally acceptable as the nineteenth century went on. In the first decades of the century, while showmanship was certainly central to the natural philosopher's activities—take the careers of Humphry Davy or Michael Faraday as examples—that showmanship was restricted to a particular context. There was a big difference between dazzling audiences with spectacular science in the genteel setting of the Royal Institution on the one hand and pulling in the crowds at the Adelaide Gallery on the other. By the end of the century however, few eyebrows would have been raised by Lord Kelvin's activities as a juror at International Exhibitions. Science had ceased to be a gentlemanly vocation and become a hard-nosed profession. Science and electricity at these massive fin de siècle scientific gatherings were weapons in the cause of imperialist expansion. Exhibitions from the Crystal Palace onwards were the occasion for a great deal of rhetoric concerning their role in establishing international harmony, mutual understanding, and peaceful commerce. In reality, however, their internationalism had a hard competitive edge. These were occasions for the ostentatious display of commercial, technological, and scientific supremacy. The heated discussions between German and British delegates at the 1881 International Electrical Congress organized at the Paris Exhibition that year concerning the introduction of absolute standards of measurement in electricity are—as we shall see in the final chapter—a good indication of the extent to which that national competitiveness could be found at the very core of scientific culture. Electricity's very visibility and the way in which it increasingly permeated Victorian culture made its disciplining increasingly crucial.

5

# The Science of Work

As the nineteenth century progressed, the concept of work was becoming a focus of attention to an unprecedented degree. More than an occupation, a means of earning a living, or even an indicator of social status, work was increasingly regarded as a moral imperative as much as a physical necessity. Work was treated as a measure of a person's moral worth in just the same way as it was an indicator of an individual's economic value. The key to this in many ways was of course the Industrial Revolution. Massive transformations were taking place in the organization of labour in factories and workplaces. New machines and processes were deskilling workers, introducing new categories of labor, and changing perceptions of what it meant to work. Andrew Ure, the Scottish chemist and enthusiast for the new factory system—"the Pindar of the automatic factory,"[1] as Karl Marx called him—remarked gleefully that the very meaning of the word "manufactory" (later shortened to the now-familiar "factory") had been turned on its head by the progress of industry: "The term Manufacture, in its original signification, undoubtedly means a work performed by hand; whereas at present it almost signifies a work performed *without* hands."[2] The machine seemed to be taking over in the world of work. The new

---

[1] K. Marx, *Capital* (1867; reprint, Harmondsworth: Penguin Books, 1976), 544.

[2] A. Ure, *The Philosophy of Manufactures* (London, 1835), 1.

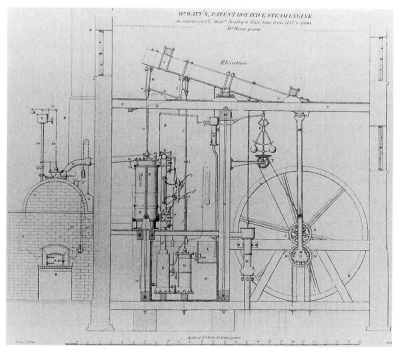

**5.1** James Watt's steam engine. Trying to understand the relationship between heat and
motion in engines such as this would play an important role in the development of
nineteenth-century thermodynamics.

science of political economy, drawing largely on Adam Smith's hugely
influential *Wealth of Nations*, published in 1776, sought to make sense of
these transformations and particularly of the new role machinery seemed
to be taking on at the center of production. The focus of attention was the
steam engine, hailed as an unparalleled new source of productive power.

The hero of steam was the Scotsman James Watt (figure 5.1). He was
not, of course, the actual inventor of the steam engine. Inventors such as
Thomas Savery and Thomas Newcomen had already applied the power
of steam to practical use, using their engines to pump water out of mines.
Watt was celebrated, however, for the improvements he had carried out on
the steam engine, significantly increasing its efficiency with his invention
of the separate condenser. He was the archetypal self-made man to be admired for the way in which he had put his own ingenuity and knowledge
to work. Originally a Glasgow instrument maker in the 1760s, he was
seen as having put his links to natural philosophers and chemists such as
Joseph Black and William Cullen, both experts on the science of heat, to

good use. The rapid expansion in use of the steam engine following Watt's innovations drew increasing philosophical interest in its workings. Men of science eager to demonstrate the practical utility of their vocations speculated on its operations. The science of heat increasingly became the focus of intense philosophical concern. Some natural philosophers argued that heat was an imponderable fluid like light or electricity. It was the presence or absence of this fluid—called "caloric"—that made a body hot or cold. Others suggested that heat should be regarded as vibrations in the particles that made up a body. The more motion was imparted to these particles, the hotter the body appeared to be.

The steam engine was widely recognized as having played a major role in bringing about Great Britain's industrial supremacy by the end of the eighteenth century. To Britain's enemies and industrial rivals, conquering the steam engine seemed the key to conquering the nation as well. Revolutionary France, in particular, regarded the systematic application of science to industry as being as much a prerequisite of supremacy as was the valor of its soldiers. One product of the Revolution in France was the systematic training of engineers and men of science. The prestigious École Polytechnique was devoted to producing cadres of trained men able to put their scientific and technical skills at the service of the state. Even following Napoleon's defeat this focus on scientific education continued. Sadi Carnot, the son of a hero of the revolutionary wars and a pioneer of the new science of work, was a direct product of this kind of training. His work combined political economy and scientific acumen to produce a new theory of the working of the steam engine and the best means to increase its efficiency and place it at the service of France.

Efficiency was the goal for a new generation of English and Scottish natural philosophers as well. Men such as James Prescott Joule in Manchester and the Thomson brothers, James and William, in Glasgow came from heavily industrial backgrounds. Their scientific values, like their moral and religious values, were the products of booming commercial cities where a premium was placed on hard work and an eye for making the most from every shilling. In scientific terms that meant finding and defining the conditions under which engines of all kinds worked best— how to maximize their outputs for a minimum outlay in terms of fuel and labor. Natural philosophers from hardheaded industrial backgrounds wanted to know how to minimize waste in nature as well as the factory. The new science of heat—thermodynamics—forged out of British industrial culture during the 1840s and 1850s was not just about understanding the steam engine, which was increasingly regarded as the model for the way in which nature worked. William Thomson, knighted and eventually elevated to the peerage for his services to industry, used Carnot's theories and Joule's experiments to make the science of thermodynamics the exemplar of a whole new way of doing physics. Thermodynamics could demonstrate how the Universe would end in heat death. It could also be used to pour scorn on the claims of geologists and evolutionary theorists concerning the development of life on earth.

New ways of doing physics were emerging in the German lands as well. German natural philosophers of the 1840s were turning their backs on the speculative *Naturphilosophie* of the previous generation. German science as it stood at the end of the eighteenth century and the first few decades of the nineteenth was widely regarded as having become bogged down in metaphysics and rampant, unsupported speculation. A new generation of Young Turks aimed to revitalize German science and make their own careers at the same time, by refounding their disciplines on a secure foundation of empiricism, materialism, and rationalism. New scientific institutions proliferated along with new opportunities to promulgate new visions of nature and the best ways of uncovering its secrets. This was an atmosphere in which ambitious young men of science such as Hermann von Helmholtz and Rudolf Clausius could put forward grand new generalizations founded on a new vogue for careful experimentation. The new generation of German men of physics prided themselves that however grandiose their theories, they were carefully grounded in sober reality. Like their French, English, and Scottish counterparts, they sought to take the steam engine and turn it into a model of the universe.

The dynamical theory of heat as it was developed during the course of the nineteenth century posited a central role to the man of science in the development of Victorian culture. Scientific culture, according to the vision both of nature and society put forward by the pioneers of this new science, was industrial culture as well. The steam engine and its offshoots provided the force that powered Victorian society. According to the confident and hard-nosed natural philosophers who espoused the new physics of work, something very much like it provided the powerhouse on which the cosmos ran as well. The universe operated on the same principles as those that governed, or at the very least ought to govern, the well-regulated Victorian factory. Such a synthesis placed the physicist at the center of the action. He understood the dynamics of nature and of society and could therefore be trusted to oversee their operations. Across Europe and America during the nineteenth century, as men of science sought to forge careers for themselves and a secure cultural niche for the disciplines they were in the process of constructing, this strategy was played out over and over again. Constructing the science of work was part of the process of constructing the Victorian physicist in relation to his culture.

## Engineering France

Even before the French Revolution, engineering education was already playing an increasingly important role in the thinking of French state officials, particularly in the military. Following the upheavals of the last two decades of the eighteenth century, major reforms took place within the French educational system, just as they did in French society more generally. These educational reforms were implemented with a view to putting in place a highly centralized regime of education, designed to produce highly technically proficient military officers fit to serve in the modern "grande armée." The centerpiece of this new technical educational system was the École Polytechnique. Potential members of the officer class were trained there in the latest developments in science as an essential element in acquiring the knowledge necessary to engage in modern warfare. Graduates of the école, as well as being trained for the military, were prepared for a wide range of public services. Under the leadership of revolutionary pioneers such as Gaspard Monge, the école was conceived as an institution devoted to the widest possible dissemination of technical knowledge. Knowledge gained there could be put to work in improving the state of French industry, much as it might be employed to build bridges or improve military ballistics.

One figure in particular provides an ideal example of the close links in France during this period between revolutionary politics, the military, and industrial organization. Lazare Carnot—a member of the Committee of Public Safety under that architect of the Terror, Robespierre—was a hero of the Revolution; he later became one of the most prominent of Napoleon's generals. Carnot was deeply engaged in technical matters as well. He was one of the leading figures behind the École Polytechnique. He played a crucial role in efforts to introduce the division of labor and new machinery into French industry during the closing years of the eighteenth century. The aim was quite explicit. Revolutionary France badly needed large quantities of armaments to fight a war on several fronts. At the same time, contemporary commentators were well aware of the military advantages Britain gained from its expanding industries. Exporting the Revolution was going to mean importing British industrial technologies such as the steam engine and new means of organizing the workforce. Carnot was an engineer and a savant in his own right as well. His *Essai sur les Machines en General* of 1783 and *Principes Fondementaux de l'Equilibre et du Mouvement* of 1803 offered a general theory of the working of machines. He was particularly known for his researches on the work done by waterwheels, linking the work done by the turning wheel to the fall of water between two different levels.

Lazare's son, Sadi Carnot, born in 1796, was therefore well placed to develop an interest in work and engines and to recognize their importance to the French Republic. Sadi was educated by his father until the age of sixteen, when he entered the prestigious École Polytechnique in 1812. By the time he was approaching graduation in 1814, however, Paris was under siege and the Napoleonic regime was rapidly coming to an end. He joined the Corps of Engineers and remained an army officer for most of the rest of his life. Following the restoration of the French monarchy, Lazare Carnot was exiled and Sadi Carnot found himself laboring under the stigma of a now infamous family name. In 1820 he was retired on half pay; he settled in Paris, where he moved on the fringes of philosophical circles, attending lectures at the Sorbonne, the École des Mines, and elsewhere. He played a leading role in the Association Polytechnique, a society of former students from the École Polytechnique with an interest in popular scientific education. Barred from elevation in the ranks or from public office as a result of his unfortunate family connections, he had plenty of time on his hands.

When Sadi Carnot visited his father in exile, Lazare told his son: "If real mathematicians were to take up economics and apply experimental methods, a new science would be created—a science which would only need to be animated by the love of humanity in order to transform government."[3] Carnot spent much of his time traveling through France and the rest of Europe, taking advantage of his opportunities to study industrial organization and practical economics by visiting factories. He made himself expert on the economics of industrializing society and on the machines that operated Europe's factories. The result of all this, coupled with his interest in popular scientific education, was a small pamphlet, *Reflexions sur la Puissance Motrice du Feu*, published in 1824. The book was about heat engines, or more particularly, the steam engine, which already seemed "destined to produce a great revolution in the civilized world" and would "afford to the industrial arts a range the extent of

[3]Quoted in S. Carnot, *Reflections on the Motive Power of Fire*, ed. E. Mendoza, xii.

which can scarcely be predicted."[4] Carnot wanted to analyze the heat engine—to find out how it operated, how heat produced work, and what the limits of its efficiency might be. This was to be his contribution to applying the experimental method to economics.

Carnot regarded heat as central to the operations of the natural economy. Nature was an "immense reservoir" of heat that could be regarded as responsible for a whole range of phenomena, from the agitations of the atmosphere to earthquakes and volcanic eruptions. To understand the ways in which heat acted to produce such different kinds of movements, Carnot argued that the problem needed to be expressed in as general a way as possible. The problem with previous analyses was that they had been too specific—too wedded to particular mechanisms. The result was that it became difficult to recognize the general laws and principles underlying the phenomena. The mechanical theory—Newton's laws of motion—could be used to analyze those machines that were put in motion by sources of power other than heat. In those cases "all imaginable movements are referred to these general principles, firmly established, and applicable under all circumstances." This was what was wanted for heat engines as well. "We shall have it," argued Carnot, "only when the laws of physics shall be extended enough, generalized enough, to make known beforehand all the effects of heat acting in a determined manner on any body."[5]

Carnot started, nevertheless, with a specific analysis, following the steam engine through its cycle of operations. "What happens in fact in a steam-engine actually in motion? The caloric developed in the furnace by the effect of the combustion traverses the walls of the boiler, produces steam, and in some way incorporates itself with it. The latter carrying it away, takes it first into the cylinder, where it performs some function, and from thence into the condenser, where it is liquefied by contact with the cold water which it encounters there. Then, as a final result, the cold water of the condenser takes possession of the caloric developed by the combustion. It is heated by the intervention of the steam as if it had been placed directly over the furnace. The steam is here only a means of transporting the caloric. It fills the same office as in the heating of baths by steam, except that in this case its motion is rendered useful."[6] This was the crucial fact for Carnot. What mattered in a steam engine—and in any other kind of heat engine for that matter—was the movement of caloric from a hot to a cold body, not its consumption. It was this movement

[4]Ibid., 3–4.    [5]Ibid., 6.    [6]Ibid., 6–7.

that produced work. "The production of motive power is then due in steam-engines not to an actual consumption of caloric, but *to its transportation from a warm body to a cold body*, that is, to its reestablishment of equilibrium—an equilibrium considered as destroyed by any cause whatever, by chemical action such as combustion, or by any other."[7]

Carnot's conclusion was that "[w]herever there exists a difference of temperature, wherever it has been possible for the equilibrium of the caloric to be re-established, it is possible to have also the production of impelling power."[8] Indeed, work could be carried out only if there were a difference in temperature. If all the Universe were at the same temperature, there could be no flow of caloric and consequently no work could be performed. The question then arose of the exact relationship between work and heat: "Is the motive power of heat invariable in quantity, or does it vary with the agent employed to realize it as an intermediary substance, selected as the subject of the action of the heat?"[9] Carnot had established that a temperature difference allowed work to be done; he also argued that the situation could be reversed: "wherever we can consume this power, it is also possible to produce a difference of temperature, it is possible to occasion destruction of equilibrium in the caloric."[10] The question he aimed to answer was whether the amount of work produced by the flow of caloric from a higher to a lower temperature was the same in all cases as the amount of work required to produce that difference of temperature in the first place. There was a distinct analogy between Sadi Carnot's model of caloric doing work by flowing from one temperature level to another lower one and his father Lazare's analysis of the work done by water turning a waterwheel while flowing from a higher level to a lower.

Sadi Carnot argued that the motive power produced by the fall of caloric from one temperature level to another could never exceed the amount of work that would be required to produce that temperature difference. Otherwise perpetual motion would be possible—a proposition that he, like all respectable late eighteenth- and nineteenth-century natural philosophers, regarded as absurd. The bulk of his pamphlet was devoted to detailed calculations and illustrations to demonstrate the general principle that "[t]he motive power of heat is independent of the agents employed to realize it; its quantity is fixed solely by the temperatures of

---

[7]Ibid., 7.    [9]Ibid., 9.
[9]Ibid., 8.    [10]Ibid.

the bodies between which is effected, finally, the transfer of the caloric."[11] He was emphatic, nevertheless, that the ideal would never in practice be attainable. The point of his work was in the end both pragmatic and socially progressive—to put the theory of how to produce work from heat at the disposal of France. "To know how to appreciate in each case, at their true value, the considerations of convenience and economy which may present themselves; to know how to discern the more important of those which are only secondary; to balance them properly against each other, in order to attain the best results by the simplest means: such should be the leading characteristics of the man called to direct, to co-ordinate the labours of his fellow men, to make them co-operate towards a useful end, whatsoever it may be."[12]

Carnot's treatise was formally presented to the French Académie des Sciences by his friend the engineer and veteran of Napoleon's Egyptian campaign, Pierre Simon Girard. It was not well received. Few members of the elite academy regarded Carnot's *Reflexions* as being in the least germane to their concerns. Girard made other efforts to bring his friend's work to public attention as well. Writing in the prorepublican *Revue Encyclopédique* he averred that "Monsieur Carnot is not afraid of tackling difficult questions; and in this first production he shows himself capable of going into a matter which has become today one of the most important with which theoreticians and physicists can occupy themselves."[13] Among engineers, anxious to exploit his findings concerning ways of maximizing the efficiency of heat engines of various kinds, Carnot's work was better thought of. Academicians, however, regarded his arguments as long-winded, poorly formulated, and frequently incorrect. Carnot's work disappeared almost without trace, particularly following his early death from cholera in 1832. It was hardly surprising that when William Thomson, visiting Paris two decades after its publication, searched high and low for a copy of the *Reflexions*, he could not find a copy, nor even a bookseller who had ever heard of it.

One Frenchman who did pay attention to Carnot's work, however, was the engineer and mathematician Émile Clapeyron. Unsurprisingly, Clapeyron shared many of Carnot's broader interests and concerns. He was a graduate of the prestigious École Polytechnique who had published extensively on the organization of public projects and popular technical education, as well as on technical engineering matters. He had worked in Russia with the French engineer Gabriel Lamé during the 1820s before

[11] Ibid., 20.    [12] Ibid., 59.    [13] Ibid., xiii.

returning to France in 1830. In 1834, he published a paper in the *Journal de l'École Polytechnique* entitled "Memoir on the Motive Power of Heat." Here he set out essentially to translate Carnot's work into the abstract mathematical language familiar to the French academic elite. Intriguingly, as well, he expressed Carnot's findings in terms of indicator diagrams: effectively graphs of pressure against volume. They had originally been developed by James Watt as a way of determining the work done by his improved steam engines and were a closely guarded trade secret. Clapeyron had possibly encountered them in Russia, where they were used by Boulton and Watt engineers building steam engines for Russian manufacturers under license. They provided in any case a striking and easily accessible way of representing Carnot's theories.

Studying and improving the efficiency of steam engines was certainly a matter of ongoing concern among French engineers, anxious to find ways of placing their industries on a par with those of their old enemies across the English Channel. The steam engineer Marc Séguin's *De l'Influence des Chemins de Fer* (1839) discussed matters pertaining to steam engine efficiency at some length. Later in the century, Séguin was to point to this publication as containing an early expression of the principle of the equivalence of heat and work. A mark of the importance increasingly accorded such matters by the French government was the award by the Ministry of Public Works of financial support to Victor Regnault—one of the rising stars of French physics during the 1830s—to carry out extensive experimental research on steam engine efficiency. Regnault embarked on a systematic effort to redetermine experimentally all the data that might be required. The task on which he set out was enormous. The results of his endeavors were not published in full until 1870. In the meantime, however, Regnault's Paris laboratory became one of the premier sites in Europe for the experimental investigation of the science of work.

French concern with the science of work, like French science more generally, largely revolved around the state. During the revolutionary and Napoleonic wars and later in the century in terms of commercial rivalry, finding ways of improving the productivity of French industry were seen as an important national imperative. Not only national economic productivity, but national pride—and during the wars, national survival as well—depended on putting science at the state's service. Sadi Carnot's work, straddling the permeable boundary between economics and physics, was an effort to bring abstruse natural philosophy to bear on a very practical question—how to make steam engines work better.

The aim of such improvement was to enhance economic productivity, to increase the amount of work available, and in the end enhance the well-being of mankind. In that sense at least, Sadi Carnot's *Reflexions* was an eminently Republican text. It was meant as an example of the way in which physics could be a matter of real, tangible public benefit. It was also firmly technocratic in its vision. The answer to France's (and society's) ills lay in the hands of technical experts, prepared and willing to put their knowledge of nature at the service of the state.

## The Culture of Dissipation

Britain in the early decades of the nineteenth century was a rapidly industrializing country. Towns that had been little more than villages a few decades earlier were in the process of being transformed into sprawling cities, their populations booming with the creation and expansion of new industries. In cotton mills, factories, and mines throughout the country, the steam engine and the division of labor were proving the foundation of unprecedented economic expansion. A new culture was being forged as well in these industrial towns and cities. The self-made men who ruled the roost in these new urban conglomerations were a very different breed from the old aristocracy and professions that still dominated the metropolis and the country's political life. Having pulled themselves up by their bootstraps, they valued self-help and hard work. They abhorred waste and inefficiency. The new industrialists regarded the single-minded pursuit of wealth as a perfectly respectable goal in its own right. Their politics was largely liberal and free-market, their religion nonconformist, though as the century progressed they veered more and more towards Toryism and the Anglican Church. Many of this new breed admired science, as long as it could be put to practical use for the benefit of mankind and the size of their purses. In particular, the science they valued also valued work, efficiency, and the diminution of waste.

Manchester was a major center for this new self-confident provincial culture. Its population of more than 300,000 in the 1830s was more than ten times its size a century earlier. It was an important center for the cotton industry that produced so much of the country's newfound manufacturing strength. It also boasted a flourishing Literary and Philosophical Society, founded as early as 1781 as a forum through which to express the cultural and philosophical aspirations of its rising middle classes. When James Prescott Joule, a successful brewer's son from Salford, a suburb of Manchester, started experimenting on engine efficiency in the late 1830s, the city already had a thriving scientific culture. Joule's tutor was the eminent chemist John Dalton, and he was soon collaborating with the electrician William Sturgeon, recently moved from London to be superintendent of the Royal Victoria Gallery for the Encouragement and Illustration of Practical Science. Joule's early experiments were on the newly invented electromagnetic engines. As we saw previously, however, he rapidly became discouraged with this work and expanded his researches to look at the question of engine efficiency more generally. He was soon fascinated by the relationship between heat and work and in trying to find ways of maximizing the production of motive power from heat.

Joule experimented diligently throughout the first half of the 1840s. His particular concern was to find ways of quantifying the relationship between heat and work—the mechanical equivalent of heat, as he called it. He argued that his experiments were conclusive proof that heat was not a particular substance—caloric—but a form of motion, or *vis viva*. With his brewing background, Joule was well versed in the precision thermometric measurements required to ground such ambitious claims. His potential audience at the Royal Society, however, was reluctant to accept his experimental efforts, rejecting his work for publication in their *Philosophical Transactions*. They did not trust this provincial, whose links were with Sturgeon and the unsavory London Electrical Society rather than with elite gentlemen of science. Joule persevered throughout the 1840s, refining his techniques and his experiments. In 1845 he presented the annual meeting of the British Association for the Advancement of Science with the results of what is now known as his paddle wheel experiment (figure 5.2). In this experiment, weights attached through pulleys to a paddle wheel enclosed in a container of water caused the paddle wheel to rotate as the weights fell. Joule argued that the experiment showed how "the force spent in revolving the paddle-wheel produced a certain increment in the temperature of the water." In other words, the motion of the weights was transformed into heat in the water. This conversion could be accurately measured: "when the temperature of a pound of water is increased by one degree of Fahrenheit's scale, an amount of *vis viva* is communicated to it equal to that acquired by a weight of 890 pounds after falling from the altitude of one foot."[14] Once again, little attention was paid to Joule's work on this occasion, but when he presented a new

---

[14]J. P. Joule, "On the Mechanical Equivalent of Heat," *Reports of the British Association for the Advancement of Science*, 1845, 15: 31.

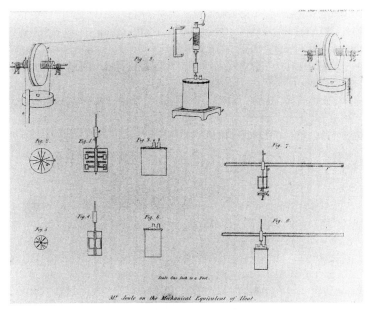

5.2 James Prescott Joule's famous paddle wheel experiments, demonstrating the mechanical equivalent of heat.

version of his experiments to the BAAS two years later in Oxford in 1847, the audience contained a very receptive pair of ears belonging to the ambitious young Glaswegian natural philosopher, William Thomson.

Like Manchester, Glasgow in the first half of the nineteenth century was a thriving industrial citadel, profiting both from the cotton mills in its hinterland and the shipbuilding industry of the city itself. Like Manchester also, it had an active scientific culture, based around the university and the Glasgow Philosophical Society, where hardheaded industrialists, businessmen, engineers, and sympathetic university academics rubbed shoulders and shared values based on work and efficiency. The Thomson family had moved to Glasgow from Belfast—another center of industry and self-help Presbyterian values—when James Thomson, the father, was appointed professor of mathematics at Glasgow College in 1832. Like their father, William Thomson and his older brother James shared in the Glasgow ethos, inherited from the eighteenth-century Scottish Enlightenment, of hard work, religious toleration, and self-discipline. James Thomson the younger, after his graduation from Glasgow College in 1840, aimed at a career in engineering, having been a passionate inventor in his youth, trying to find ways of minimizing the waste and maximizing the economic efficiency of engines. In 1843 he was apprenticed to William Fairbairn's machine-making engineering firm. William Thomson, with his talent for mathematics, on the other hand, following his own graduation from Glasgow a year later, headed for a second degree in Cambridge and the mathematical Tripos.

As we have already seen, Cambridge by the 1840s was probably one of the best places in Europe at which to acquire a rigorous mathematical training. The analytical revolution of two decades earlier had transformed the syllabus. The system of tutors guaranteed personal attention and plenty of opportunity for practice. Thomson's tutor was William Hopkins of St. Peter's College, known for his reliable production of high wranglers. Thomson did well, graduating second wrangler and first Smith's prizeman in 1845. Following his graduation, he visited Paris, where as well as searching unsuccessfully for Carnot's *Reflexions* (and making do with Clapeyron instead) he worked with Victor Regnault in his physics laboratory. Given Regnault's concern with steam engines, this was ideal further training for a hopeful natural philosopher with Thomson's interests in work, efficiency, and the annihilation of waste. The experience provided him with a thorough grounding in experimental practice to complement the mathematical expertise he had acquired at Cambridge. The experience was soon to be put into practice when later the same year he was appointed to the vacant chair of natural philosophy at Glasgow (figure 5.3). Throughout his time at Cambridge and after, William and his brother had been in constant correspondence over the vexed issue of heat, work, and waste. The issue was to vex Thomson for several decades to come.

James Thomson had encountered the Carnot-Clapeyron theory of the motive power of heat during his engineering training. As he explained it to his brother, he had learned that "during the passage of heat from a given state of intensity to a given state of diffusion a certain quantity of mechanical effect is given out whatever gaseous substances are acted on, and that no more can be given out when it acts on liquids or solids."[15] What concerned both brothers was the waste involved in the process in practice. A water mill wasted work by spillage of water from the wheel buckets; a steam engine wasted work by releasing steam or water when it was still hotter than its surroundings. Following Carnot, they both believed that caloric, or heat, was conserved during the production of work. Their problem was in how to get the most out of it. Increasingly, they

[15]Quoted in C. Smith, *The Science of Energy*, 42.

5.3 William Thomson's teaching laboratory at Glasgow. The three windows on
    the left on the ground floor are those of the physical laboratory; the three
    above belong to the apparatus room. Entrance to both rooms was through
    the door in the pentagonal tower on the left.

regarded the Carnot-Clapeyron theory, in which work was the result of
caloric flowing from one temperature to a lower one, as the best available
account of the workings of engines. William Thomson's experiments in
Glasgow on a Stirling air engine seemed to confirm this view. This was
why he found Joule's claim at the Oxford meeting of the BAAS in 1847,
of experimental proof that mechanical work was the result of an absolute
loss of heat, deeply puzzling.

    Thomson expressed his conundrum in trying to reconcile Carnot
and Joule in a famous footnote to his "Account of Carnot's Theory of the

Motive Power of Heat," first read to the Royal Society of Edinburgh in 1849: "When 'thermal agency' is thus spent in conducting heat through a solid, what becomes of the mechanical effect which it might produce? Nothing can be lost in the operations of nature—no energy can be destroyed. What effect then is produced in place of the mechanical effect which is lost? A perfect theory of heat imperatively demands an answer to this question; yet no answer can be given in the present state of science."[16] If work were simply the result of heat falling from one temperature level to another, as Carnot suggested, then what happened to the work that might have been produced if there was no engine there for it to operate on? Conversely, if, as Joule argued, the production of work required an absolute loss of heat, where did the heat go in cases where no useful work was apparently being done, as in the case of simple heat conduction? In 1849, when Thomson presented his account of Carnot's theory to the Royal Society of Edinburgh, he still had no answer to these pressing questions.

Within a few years, however, Thomson thought he had solved the problem. In a series of papers entitled "On the Dynamical Theory of Heat," presented to the Royal Society of Edinburgh between 1851 and 1855, he laid the framework of the new science of heat he would call thermodynamics. His theory rested on two central propositions. The first one, derived from Joule, stated: "When equal quantities of mechanical effect are produced by any means whatever from purely thermal sources, or lost in purely thermal effects, equal quantities of heat are put out of existence or are generated."[17] In other words, he had fully accepted Joule's assertion of the mutual convertibility of heat and work. Thomson's second proposition rested on his reading of Carnot. He stated: "If an engine be such that, when it is worked backwards, the physical and mechanical agencies in every part of its motions are all reversed, it produces as much mechanical effect as can be produced by any thermo-dynamic engine, with the same temperatures of source and refrigerator, from a given quantity of heat."[18] He had abandoned his earlier commitment to Carnot's insistence that caloric, or heat, be conserved during this process. He concluded that in any process of heat transfer that did not fulfill Carnot's criterion of perfect reversibility—in other words, in any real engine—there was "an absolute loss of mechanical energy available to man."[19]

[16]Quoted ibid., 94.     [18]Quoted ibid., 110.
[19]Quoted ibid., 107.     [19]Quoted ibid., 124.

Thomson's universe now had a sense of direction. It was a universe in which the capacity for doing useful work was continually being lost. Carnot's perfectly reversible engine was an ideal that could not exist in the real world. Every time a real engine operated, friction, bad lubrication, imperfect insulation, and any number of other effects colluded to dissipate energy. These were not just facts of life, they were facts of the Universe. Any natural process involved an inevitable loss of heat that

TREATISE

ON

# NATURAL PHILOSOPHY

BY

SIR WILLIAM THOMSON, LL.D., D.C.L., F.R.S.,

PROFESSOR OF NATURAL PHILOSOPHY IN THE UNIVERSITY OF GLASGOW,
FELLOW OF ST PETER'S COLLEGE, CAMBRIDGE,

AND

PETER GUTHRIE TAIT, M.A.,

PROFESSOR OF NATURAL PHILOSOPHY IN THE UNIVERSITY OF EDINBURGH,
FORMERLY FELLOW OF ST PETER'S COLLEGE, CAMBRIDGE.

PART I.

*NEW EDITION.*

CAMBRIDGE:

AT THE UNIVERSITY PRESS.

1890

*[All Rights reserved.]*

5.4 The title page of William Thomson and Peter Guthrie Tait, *Treatise on Natural Philosophy*.

might have been converted into useful work by falling between two temperature levels. Once that capacity to do work had been lost it could not be recovered. There was a strong moral imperative implicit in all of this. If work was continuously being lost, it behooved the canny operator to take advantage of as much of it as he could while it remained available. This was the worldview shared by Thomson, Joule, and an increasing number of other engineers and natural philosophers from nonconformist, industrial backgrounds. Men such as W. J. Macquorn Rankine, a Clydeside engineer and later professor of civil engineering and mechanics at Glasgow, and Peter Guthrie Tait (with whom Thomson wrote *Treatise on Natural Philosophy*), Cambridge wrangler from an Edinburgh banking family and professor of mathematics at Queen's College Belfast, before returning to Edinburgh as professor of natural philosophy in 1860, played key roles in helping Thomson during the 1850s and 1860s to further articulate this new science of thermodynamics, based on the mechanical equivalent of heat and the universal tendency towards dissipation (figure 5.4).

For Thomson and his coworkers in Britain, one of the major virtues of the new science of thermodynamics was precisely the way in which it provided the universe with a sense of direction. Man could direct the operations of nature but could not reverse them. The Universe had a beginning and—more importantly—it had an end. There would come a time when all heat had been dissipated. The energy would all still be there— nothing would have been lost—but the universe would be at a uniform temperature, and without heat transfer from higher to lower temperatures no work could be done and the universe would be at a standstill. This "heat death of the Universe" evoked powerful images (figure 5.5). H. G. Wells made good use of the scenario in the closing pages of *The Time Machine*. Moving forward through time, Wells's traveler eventually arrived at a desolate future: "The sky was no longer blue. North-eastward it was inky black, and out of the blackness shone brightly and steadily the pale white stars. Overhead it was a deep Indian red and starless, and south-eastward it grew brighter to a glowing scarlet where, cut by the horizon, lay the huge hull of the sun, red and motionless."[20] It was a scene that never changed, since the Earth had long since ceased rotating on its axis. Moving another thirty million years into the future, the traveler witnessed an eclipse of the sun over a now lifeless planet: "The darkness grew apace; a cold wind began to blow in freshening gusts from the east, and the showering white flakes in the air increased in number.

[20]H. G. Wells, *The Time Machine* (1895; reprint, London: Everyman Library, 1995), 73.

5.5 The end of the world and the heat death of the Universe as envisaged by Camille
   Flammarion in 1893.

From the edge of the sea came a ripple and whisper. Beyond these lifeless sounds the world was silent. Silent? It would be hard to convey the
stillness of it. All the sounds of man, the bleating of sheep, the cries
of birds, the hum of insects, the stir that makes the background of our
lives—all that was over."[21] It was a graphic description of Thomson's
universe, though the materialist Wells provided his book with a cheeky
evolutionist narrative that would have left Thomson fuming.

Indeed Thomson, and his antimaterialist allies regarded thermodynamics as a powerful weapon with which to counter Darwinian evolution. In a series of papers in the early 1860s, appearing shortly after the
publication of Darwin's *Origin of Species* in 1859, Thomson argued that
thermodynamics clearly demonstrated that natural selection was wrong.
The Sun could not have been provided with an infinite store of energy
nor was there any obvious chemical or mechanical source of energy with
which it could be replenished. The Sun's heat had therefore to be finite.
There were only two possible hypotheses concerning the sun's origins:
"The sun must . . . either have been created as an active source of heat
at some time of not immeasurable antiquity, by an over-ruling decree;
or the heat which he has already radiated away, and that which he still

[21] Ibid., 75.

possesses, must have been acquired by a natural process, following permanently established laws. Without pronouncing the former supposition to be essentially incredible, we may safely say that it is in the highest degree improbable, if we can show the latter to be not contradictory to known physical laws. And we can do this and more, by merely pointing to certain actions, going on before us at present, which, if sufficiently abundant at some past time, must have given the sun heat enough to account for all we know of his past radiation and present temperature."[22] Thomson's calculations and his figures provided him with a conservative estimate of the age of the sun: "It seems, therefore, on the whole most probable that the sun has not illuminated the earth for 100,000,000 years, and almost certain that he has not done so for 500,000,000 years. As for the future, we may say, with equal certainty, that inhabitants of the earth cannot continue to enjoy the light and heat essential to their life, for many million years longer, unless sources now unknown to us are prepared in the great storehouse of nature."[23] Similar calculations placed strict limits on the possible age of the Earth, making Darwin's estimate of 300,000,000 years for the "denudation of the Weald" alone, for example, appear hopelessly optimistic.

The dynamical theory of heat, in Thomson's articulation, was very much a product of a Scottish and northern English industrial sensibility. Its creators came from backgrounds that valued hard work, self-discipline, and thrift and abhorred dissipation and waste. This is what the dynamical theory was about. Waste was endemic—an unavoidable feature of life built into the very fabric of the universe. Waste could be minimized, however, even if could not be eliminated entirely, by careful and disciplined attention. It was a theory that arose from very pragmatic considerations. Joule's initial interest in engine efficiency had come about as a result of his experiments to maximize the efficiency of the electromagnetic engines enthused over by his fellow electricians in the London Electrical Society. His concerns about the practical improvement of engines led him to consider the relative advantages of work from different sources. Thomson had been motivated by the equally practical concern he shared with his brother to find a theory that could be exploited to minimize the waste of power in steam engines. These concerns mattered to them because they were concerns that mattered to the cultures they inhabited. They wanted to put natural philosophy to work so that it

[22]W. Thomson, *Popular Lectures and Addresses* (London, 1889), 1: 363–64.
[23]Ibid., 1: 368.

could minimize the wastage of the industrial society they inhabited and increase its profits. The thermodynamic model also had the virtue of recapturing the idea of progression from those they regarded as dangerous materialists.

## The German Science

As we saw earlier, major changes were afoot in German natural philosophy during the second quarter or so of the nineteenth century. A new generation of practitioners were anxious to dissociate themselves from what they regarded as the metaphysical excesses of the previous generation's *Naturphilosophie*. In Prussia in particular, major educational reforms during the early years of the century had greatly expanded the number of students attending university. Unlike at the old English universities, research was increasingly starting to be considered part of a professor's duties at these new German institutions. The new University of Berlin, opened in 1809, had close links with the Berlin Academy of Sciences. It was entirely funded by the Prussian state. By midcentury, the reforms were coming to fruition. The population was educated to an unprecedented degree. These were not disinterested initiatives on the part of the Prussian state. Education, and technical education in particular, was seen as the key to industrialization and modernization. This state-sponsored expansion of the educational system, in turn, provided opportunities at all levels to the new generation of hopeful natural philosophers. It provided them with the resources to reshape German science in their own image and to forge it into a central feature of nineteenth-century German notions of statehood. The science of work was crucial here as well. It was to prove an indispensable linchpin in the process of refounding German science on a materialist, rationalist (and nationalist) basis.

While his English, French, and Scottish contemporaries became engaged in constructing a science of work as a product of their concern with the steam engine and its efficiencies, the German doctor Julius Robert Mayer, as befitted his profession, was more concerned with the human body. Born the son of an apothecary in the city of Heilbronn in the kingdom of Württemberg in 1814, Mayer studied medicine at the University of Tübingen during the early 1830s. In autobiographical sketches written later in life, he maintained that as a child he had been fascinated by machinery and the quest for a perpetual motion engine. Having completed his medical degree at Tübingen by 1838, he went to Amsterdam, where, having completed the relevant examinations, he enlisted as a ship's doctor with the Dutch East India Company. After spending some time in Paris, he embarked from Rotterdam aboard the ship *Java*, sailing for the Dutch East Indies in 1840. In the course of his duties as ship's doctor, he became aware of the unusual color of the venous blood of his shipmates. It was unusually red, appearing more like arterial than venous blood. The implication was that the heat of the tropics bore some relationship to the oxygenation of the blood. He eventually concluded that in the tropical heat the body expended less effort in maintaining its internal heat and that as a result less oxidation took place in the blood. It was to this observation that he attributed his interest in heat, work, and the body.

Back in Heilbronn in 1841 and embarking on a medical practice on dry land Mayer tried to interest others in his speculations concerning heat and work. He tried unsuccessfully to publish his work in the prestigious *Annalen der Physik und Chemie*, edited by Johann Christian Poggendorff. In his first published work, "Remarks on the Forces of Inanimate Nature," published in the *Annalen der Chemie und Pharmacie* in 1842, he argued for a relationship between "fallforce," motion, and heat: "We can make clear to ourselves the natural connection existing between fallforce, motion and heat in the following way. We know that heat appears when the individual massy particles of a body move closer to each other; compression produces heat; now, what holds for the smallest massy particles and the smallest spaces between them must well also apply to large masses and measurable spaces. The descent of a weight is a real reduction in the volume of the earth, and thus must certainly stand in connection with the heat that thereby appears; this heat must be exactly proportional to the size of the weight and its (original) distance. From this consideration one is led quite easily to the above-discussed equation of fallforce, motion, and heat."[24] Such hypothetical arguments were unlikely to gain favor with the new leaders of German science, committed to precision in experiment and language.

Mayer attributed to his voyage on the *Java* the discovery "that motion and heat are only different manifestations of one and the same force, and that consequently motion or mechanical work and heat, which had hitherto mostly been regarded as entirely disparate things, must also be able to be converted and transformed into one another."[25] He had a specific figure for the quantitative relationship between motion and heat in mind as well, derived from published figures for the heating of air by

---

[24] Quoted in K. Caneva, *Robert Mayer and the Conservation of Energy*, 24–25.

[25] Quoted ibid., 28.

compression. He asserted that "the fall of a given weight from a height of around 365 meters corresponds to the heating of an equal weight of water from 0° to 1°."[26] Mayer's work had little impact at the time, though it presumably impressed the eminent chemist, Justus Liebig, who edited the *Annalen der Chemie und Pharmacie*. Mayer had little to say about the matters that concerned other enthusiasts for the science of work. His theories were couched in speculative and obscure terms as well. To many of his potential audience he must have read like a *Naturphilosoph* himself, although he roundly repudiated any such connection. He made clear, moreover, that he was no materialist—and materialism was very much in vogue among the new generation of German natural philosophers.

Like his near contemporary Mayer, Hermann von Helmholtz approached the science of work from a medical perspective as well. Born in 1821, the son of a Prussian *Gymnasium* teacher, Helmholtz studied medicine at the University of Berlin, with the Prussian army paying his way. In return for four years of medical education he undertook to spend the next eight years serving the army as a medical officer. He therefore spent the years between 1843 and 1848 as a staff surgeon based at Potsdam until his patron, the eminent anatomist and physiologist Johannes Müller, engineered his early discharge from the military and secured for him in 1849 a position as associate professor of physiology at the University of Königsberg. During his years in the army, Helmholtz had been a prolific publisher on physiology, having carried out experiments on, among other things, the role of heat in muscle physiology. He was part of an ambitious young coterie of experimental physiologists, including Emil du Bois Reymond and Carl Ludwig, who were anxious to turn physiology into a robustly materialist experimental science, purged of dubious metaphysical trappings like the life force. For this group of physiologists anything that happened in the body, like anything that happened in the steam engine, should be measurable and strictly quantifiable.

In 1847, Helmholtz published his latest contribution to this campaign—*Über die Erhaltung der Kraft* (On the Conservation of Force). The essay was privately published in pamphlet form. Like Mayer before him, Helmholtz had attempted to interest Poggendorff and his prestigious *Annalen der Physik*, but had also been rebuffed. Helmholtz's physiological work had been aimed at showing how the heat of animal bodies and their muscular action could be traced to the oxidation of food—their fuel. In this he was following in the footsteps of Liebig, who had pioneered

[26]Quoted ibid., 25.

research into the connections between the chemistry of nutrition and vitality. Where Carnot, Joule, and Thomson had been concerned to identify the origins of work in the steam engine, Helmholtz was concerned with its origins in the human engine. One of Helmholtz's particular concerns was to show that the supposition of a vital living force violated the principle of the impossibility of perpetual motion, since a living force would in principle be producible out of nothing. All the work produced by a human body, like all the work done by a steam engine, had to be accounted for. In his essay Helmholtz posited a purely mechanical universe in which the exact quantitative relationship of different forces to each other was susceptible of mathematical demonstration.

For Helmholtz, there was an obvious connection between human and machine work. "The idea of work is evidently transferred to machines from comparing their performances with those of men and animals, to replace which they were applied. We still reckon the work of steam-engines according to horse-power."[27] He traced the origins of the science of work to eighteenth-century efforts to produce mechanical automata that could replace living beings. This focused attention on the forces that animated living bodies and their origins. This was then the basis for comparison between humans and machines in terms of the sources of the work they performed. The conclusion was the principle of the conservation of force: "We cannot create mechanical force, but we may help ourselves from the general storehouse of Nature. The brook and the wind, which drive our mills, the forest and the coal-bed, which supply our steam-engines and warm our rooms, are to us the bearers of a small portion of the great natural supply which we draw upon for our purposes, and the actions of which we can apply as we think fit. The possessor of a mill claims the gravity of the descending rivulet, or the living force of the moving wind, as his possession. These portions of the store of Nature are what give his property its chief value."[28] Nature, from this perspective, was a storehouse of work that could be exploited through various mechanisms, including the human body and the steam engine.

While Mayer and Helmholtz were moved by physiological considerations to investigate the science of work, Helmholtz's fellow Prussian, Rudolf Clausius, approached the issue, like his British and French

[27] H. Helmholtz, "The Interaction of Natural Forces," in D. Cahan (ed.), *Science and Culture*, 20.

[28] Ibid., 29.

contemporaries, from the perspective of the steam engine. Born in 1822, he entered the University of Berlin in 1840, studying mathematics and the natural sciences. After qualifying as a *Gymnasium* teacher he remained in Berlin throughout the 1840s, obtaining a doctorate in 1848 with a dissertation on "the light-dispersing and luminous effects of the atmosphere through theoretical considerations,"[29] following in the footsteps of his patron, the physicist Heinrich Gustav Magnus. Following the completion of his doctoral studies, he turned to the study of the motion of gases and elastic bodies. It was this research that focused his attention on the problems of heat and work, through his reading of Regnault's experimental work and of Clapeyron's and others' theories. He was soon appointed as a physics teacher to the Berlin Artillery and Engineering School and as a *Privatdozent* (or tutor) at the University of Berlin. Unlike Mayer and Helmholtz, Clausius was successful in publishing his first venture into the science of work—"On the Moving Force of Heat, and the Laws Regarding the Nature of Heat That Are Deducible Therefrom"—in the *Annalen*. It appeared in 1850.

Clausius's publication was based on his reading of William Thomson's 1849 "Account of Carnot's Theory." His argument was simple: Thomson was mistaken in supposing that Carnot and Joule were necessarily at odds with each other in any crucial respect. It was possible, he argued, to reconcile Carnot's claim that work was the result of heat flowing from one temperature level to another, lower, temperature level, with Joule's assertion that work was the product of conversion from heat. The trick, he claimed, was simply to drop Carnot's assumption that heat was conserved during the process. There was no reason to suppose that the production of work by heat did not require *both* the flow of heat from one temperature level to another *and* the conversion of a certain proportion of the heat into work. As he put it, "It is not at all necessary to discard Carnot's theory entirely, a step which we certainly would find it hard to take, since it has to some extent been conspicuously verified by experience. A careful examination shows that the new method does not stand in contradiction to the essential principle of Carnot, but only to the subsidiary statement *that no heat is lost*, since in the production of work it may very well be the case that at the same time a certain quantity of heat is consumed and another quantity transferred from a hotter to a colder body, and both quantities of heat stand in a definite relation to the work that is

[29] Quoted in C. Jungnickel and R. McCormmach, *The Intellectual Mastery of Nature*, 1: 164.

done."[30] This was much the conclusion at which Thomson would arrive in his 1851 paper "On the Dynamical Theory of Heat." Indeed, Thomson's resolution of his dilemma was based on his reading of Clausius's paper.

Clausius worked on refining his theories of heat throughout the 1850s and beyond. Increasingly, as we shall see, he started making explicit links between the theory of heat and the work on gases in motion that had first drawn his attention to the problems of heat and work. Clausius was interested in the kinetic theory of gases—the idea that the large-scale properties of gases could be understood as the results of the small-scale movements of the particles, or molecules, of which the gases were made up. He understood heat to be simply an effect of the motion of such particles—hot gases were made up of fast-moving particles, colder gases were made up of slower particles. Work, then was the result of "the alteration in some way or another of the arrangement of the constituent molecules of a body."[31] In 1865, Clausius introduced a new word— "entropy"—into his version of the dynamical theory of heat. He could then reformulate the second principle of the dynamical theory of heat as the assertion that the entropy of the universe tends to a maximum.

Both Helmholtz and Clausius made their names with their contributions to the science of work. Not only were they established as key figures in German science, they were as a result of their work on heat players on the international stage as well. Their success and that of those like them in other new-forged disciplines, in attracting attention to their theories, also made German science successful. Throughout the nineteenth-century German laboratories and research institutes were increasingly attractive places of pilgrimage for young scientific acolytes keen to study at the feet of these new masters of physics. The science they produced was self-consciously abstract and rationalist. It was avowedly and deliberately the antithesis of the previous generation's wildly metaphysical *Naturphilosophie*. Arguably, the research tradition forged in mid-nineteenth-century German research institutes might well be regarded as the direct precursor of twentieth-century theoretical physics. It was a tradition that regarded mathematical theorizing about nature as an autonomous activity in its own right. It was becoming clear by the 1860s, however, that this German science that might appear to the casual observer as having so much in common with it, was also the direct antithesis of the new natural

[30]R. Clausius, "On the Motive Power of Heat, and on the Laws which can be Deduced from it for the Theory of Heat," 112.

[31]Quoted in C. Smith, *The Science of Energy*, 167.

philosophy that William Thomson and his acolytes in England and Scotland held so dear.

## The Statistical Universe

As we have just seen, one of the things that drew Clausius to study heat and the science of work was his interest in the kinetic theory of gases—the idea that gases were made up of large numbers of rapidly moving particles and that their heat was the result of these rapid movements. The idea that heat was a form of motion was not new, of course. Benjamin Thompson, the flamboyant refugee from the American War of Independence, had carried out experiments showing that heat was produced during the process of drilling a cannon bore, which he suggested showed that heat was a form of motion. Sir Humphry Davy came to the same conclusion on the basis of experiments involving the melting of ice by friction. James Prescott Joule regarded his own experiments on the mechanical equivalent of heat as having decisively established that heat was the result of motion. What interested Clausius was the question of just what kind of motion was heat. Was it the result of the internal particles making up a body vibrating, for example? Or was it the result of translational motion from one position to another? Heat might even be the result of particles rotating on their own axes.

Others, of course, had no truck with the idea that heat was a form of motion at all. The fate of two British contributors to discussions of such matters provides a good example of the view's marginality for much of the first half of the century. In 1820, John Herapath, an English journalist and mathematician, submitted a manuscript, "A Mathematical Inquiry into the Causes, Laws, and Principal Phenomena of Heat, Gases, Gravitation, &c," to be read at the Royal Society and published in its *Philosophical Transactions*. Among other things, Herapath in his manuscript offered a mathematical derivation of the ideal gas law (relating the pressure, volume, and temperature of a gas) on the basis that the heat of a gas was proportional to the internal motion of its constituent particles. The paper was rejected. Similarly, in 1845 a young tutor for the East India Company in Bombay, John James Waterston, submitted a paper to the Royal Society, "On the Physics of Media that are Composed of Free and Elastic Molecules in a State of Motion." It was dismissed by John Lubbock, one of the society's referees, as "nothing but nonsense."[32] Half a century later,

[32]Quoted in S. Brush, *The Kind of Motion we Call Heat*, 1: 140.

with the kinetic theory of gases well-established, Waterston's manuscript was discovered by Lord Rayleigh in the society's archives, dusted down, published in the *Philosophical Transactions* for 1892 and triumphantly hailed as a British precursor. Herapath's ideas were picked up by James Joule during the course of his own speculations on the origins of heat during the late 1840s. But Joule himself was at the time still a largely obscure figure, and his views attracted little attention.

Clausius's paper "Über die Art der Bewegung, welche wir Wärme nennen" (On the Kind of Motion that we call Heat) was published in the *Annalen* in 1857. In this paper Clausius argued that the heat of a gas must be made up of several kinds of motion on the part of its constituent particles. Particles in a gas must have rotational and vibrational motion as well as translational motion. The total heat of a gas must therefore be proportional to the sum of these motions. On the basis of this model, Clausius could calculate various properties of his hypothetical gas and compare them with the known properties of real gases. Clausius assumed that compared with the volume of a gas as a whole, the volume taken up by the particles themselves was infinitesimally small. He also assumed, for purposes of calculation, that all the particles moved with the same average velocity, which he calculated for different gases as being hundreds, if not thousands, of meters per second. In the face of objections that these figures must be false, since otherwise gases would diffuse far more quickly than they were known to do, Clausius soon abandoned the assumption that the volume of the particles themselves was infinitesimal. Instead he introduced the concept of the "mean free path," or the average distance that a particle could travel in a straight line before colliding with another particle.

Clausius's efforts alerted others to the possibilities of using a dynamical (or kinetic) theory of gases as a way of providing a convincing model of heat and work on a microscopic level. James Clerk Maxwell's "Illustrations of the Dynamical Theory of Gases," published in the *Philosophical Magazine* in 1860, made use of Clausius's concept of the mean free path. But where Clausius had every particle in a gas moving at the same average velocity, Maxwell drew on the science of statistics to allow for a random distribution instead. Nineteenth-century statistics had its origins in the study of populations of people rather than of inanimate objects. Maxwell had become interested in statistical theory as a student at Cambridge, after reading John Herschel's review in the *Edinburgh Review* of Adolphe Quetelet's work on probability. Quetelet was interested in what he called "social physics" and turned to statistics as a way of understanding large populations. Herschel's review was an effort to bring Quetelet's work to a wider audience and certainly seems to have convinced the young Maxwell, who wrote to his friend Lewis Campbell that "the true logic for this world is the Calculus of Probabilities... the only 'Mathematics for Practical Men', as we ought to be."[33] This was the logic that he applied about ten years later to understanding the dynamical theory of gases.

Maxwell, in his *Philosophical Magazine* contribution, argued that collisions between the constituent particles of a gas would produce a distribution of velocities rather than leading towards an equalization of velocities. On this basis, he calculated what the statistical law governing the distribution of velocities in a gas might be. He also established the equipartition theorem, arguing that the energy of the particles of a gas was equally distributed among their modes of motion: rotational, translational, and vibratory. The picture of the microscopic world that Maxwell produced was one of particles whizzing haphazardly through space, colliding with each other and bouncing off in random directions. The movements of these individual particles were impossible to predict. Just what these particles were also remained an open question. Like William Thomson, Maxwell toyed with the idea (put forward by Helmholtz) that they were in fact vortices in the ether and that the large-scale properties of matter could in theory be calculated from the characteristics of these vortices. One thing Maxwell was sure of, however. The motions of these particles might be random, but the particles themselves were not. Maxwell not only argued that particles of the same elements of matter were identical, but that their identity was proof of the existence of a divine manufacturer.

Maxwell was keenly aware that the probabilistic nature of his arguments caused problems for the new thermodynamics. In particular, it raised questions about irreversibility and the second law of thermodynamics. He illustrated the problem in a letter to P. G. Tait in 1867 by introducing the idea of an intelligent being who had the ability to change the direction of individual particles of a gas. Maxwell imagined a situation where particles of gas were confined in two partitions, separated by a frictionless sliding door. The particles of gas in each partition would have a distribution of velocities as determined by Maxwell's own distribution law. By opening and closing the door at the appropriate time, the being—"Maxwell's Demon," as William Thomson later baptized him— could change the distribution of velocities, confining the faster particles

[33]L. Campbell and W. Garnett, *The Life of James Clerk Maxwell* (London: Macmillan, 1882), 143.

to one partition and the slower ones to another. The result would be that the gas in one partition became hotter and the gas in the other became colder—an apparent flow of heat from a cold to a hot body without any work having been carried out on the system, in seeming contradiction to the second law of thermodynamics. Maxwell used the apparent paradox of the demon to raise questions about determinism and the relationship between physical phenomena on the microscopic and the macroscopic scale. What it showed, he argued was that the second law of thermodynamics "has only a statistical certainty."[34]

Maxwell's ideas about interpreting thermodynamic laws statistically were taken up by the Austrian physicist Ludwig Boltzmann. Boltzmann, educated at Linz and at Vienna under the physicist Josef Stefan, was in many ways a typical product of the new style of German physics education. He argued that the laws of thermodynamics could no longer be taken to apply absolutely in the microscopic realm. In particular the second law of thermodynamics, stating that the entropy of the universe was continually increasing, had to be understood as a statistical, rather than a strictly deterministic, generalization. In other words it was possible, if not very likely, that under certain circumstances entropy could decrease. His ideas were, to put it mildly, controversial. Boltzmann clashed repeatedly during his career with opponents such as the physicist and philosopher Ernst Mach, who argued that Boltzmann's ideas made nonsense of the whole enterprise of physics. Mach felt that Boltzmann's ideas were excessively metaphysical and went beyond the boundaries of what was observable and therefore knowable in science. Many of his German contemporaries had problems with Boltzmann's insistence on the reality of the atomic theory, which they felt went against current trends towards doing away with mechanical models in the British tradition. In Boltzmann's case these abstruse discussions had tragic consequences. Boltzmann committed suicide in 1906, apparently convinced that the world of physics had failed to understand and appreciate his ideas.

Critics such as Mach were unhappy with the identification of thermodynamics with the atomic theory. They felt that physics should deal only with observable phenomena and avoid the unnecessary introduction of theoretical entities like atoms. The American physicist Josiah Willard Gibbs shared this concern, arguing that "one is building on an insecure foundation, who rests his work on hypotheses concerning the constitution of matter." Gibbs had gained his degree and his doctorate at Yale

---

[34]Quoted in P. Harmann, *The Natural Philosophy of James Clerk Maxwell*, 139.

before spending three crucial years studying in Paris, Berlin, and Heidelberg during the 1860s. He returned to the United States to become professor of mathematical physics at Yale with a distinctly Germanic perspective on physics. Gibbs's work in thermodynamics was devoted to reducing the science to its simplest possible formulation. As he put it, "In the present state of science, it seems hardly possible to frame a dynamic theory of molecular action which shall embrace the phenomena of thermodynamics, of radiation, and of the electrical manifestations which accompany the union of atoms." Rather than trying to explain the "mysteries of nature" in this way, Gibbs declared himself "contented with the more modest aim of deducing some of the more obvious propositions relating to the statistical branch of mechanics."[35] His *Elementary Principles in Statistical Mechanics* (1902) sought to place thermodynamics on a "rational foundation" that avoided paying too much attention to elaborate mechanical models.

Maxwell's view of the implications of the atomic theory was unambiguous: "if the molecular theory of the constitution of bodies is true, all our knowledge of matter is of the statistical kind. A constituent molecule of a body has properties very different from those of the body to which it belongs . . . The smallest portion of a body which we can discern consists of a vast number of such molecules, and all that we can learn about this group of molecules is statistical information."[36] All our knowledge of the universe, in other words, was statistical. This meant that the kind of knowledge that was discoverable about the microscopic world of molecules in motion was the same kind of knowledge that was discoverable about human society—it applied to groups rather than single individuals. In fact, in a paper read to the Cambridge Apostles in 1873, "Does the progress of Physical Science tend to give any advantage to the Opinion of Necessity (or Determinism) over that of the Contingency of Events and the Freedom of the Will?," Maxwell implied that it was the statistical nature of the Universe that provided it with its sense of direction. The second law of thermodynamics did not necessarily apply to single particles. It did, however, apply to the Universe as a whole. The statistical logic that Maxwell had described as the only true logic for "practical men" really did seem to underpin that most practical of sciences, thermodynamics. Maxwell's contribution to the Apostles' debate suggests, moreover, that quite a lot hinged on what might otherwise

[35] Quoted in S. Brush, *Statistical Physics and the Atomic Theory of Matter*, 77.

[36] L. Campbell and W. Garnett, *The Life of James Clerk Maxwell* (London: Macmillan, 1882), 439.

appear to be rather esoteric discussions about the choice of appropriate mathematical method or mechanical model. Such discussions had
important ramifications for the scope and limits of human knowledge.

## Conclusion

The nineteenth-century science of work developed in different ways in
different contexts. Natural philosophers in England, France, the German
lands, and Scotland had a range of interests and concerns. They might
share what the historian and philosopher Thomas Kuhn has described
as a "concern for engines," but the various ways in which that concern
manifested itself was wholly contingent upon particular local cultures.
The Industrial Revolution in England and Scotland had already resulted
in massive and continuing changes to the cultural (and physical) landscape of the two countries. Similar processes were under way in France
and the German lands by the second quarter of the century. One outcome
of this was to focus philosophical and practical attention on the problem
of work—its origins and the ways of maximizing its output. The ways in
which this shared concern were manifested were, however, very different
from country to country. In a variety of ways the science of work proved
to be a good way of forging a scientific career as well. It was a way of
showing how natural philosophy could itself be put to work for the national good. In this way it was to prove central to the process of finding
and defining an increasingly central role for the scientist in public culture
as the nineteenth century progressed.

British natural philosophers in particular were keen to emphasize the
cosmological role of the science of work. They saw the second law of
thermodynamics in particular as playing a central role in making the
universe progressive. It was to be understood as a grand principle of
dissipation that showed how the universe was gradually progressing as
energy became dissipated and no longer available for conversion into
useful work. It was a conception of progress that could be put to work
to counter the then prevailing materialist view of progress popularized
by the *Vestiges of the Natural History of Creation*. It could be used as well
to expose the fallacies and pretensions of geologists and proponents of
evolution by natural selection, who required that the earth have an indef-
initely long history to provide the time required for their developmental
mechanisms to operate. Joule's and Thomson's physics could show them
that such an indefinite bank of time simply could not have existed. This

is an example of the centrality of local cultural concerns to the development of the science of work throughout the century. Making common ground among physicists from different cultures and backgrounds as to what the science of work really was itself required work. It could have important implications not just for understanding the steam engine, or even understanding the Universe, but for understanding the nature of knowledge itself. As disputes between German and British pioneers and their supporters demonstrated in particular, agreement as to who the discoverers of thermodynamics were first of all required agreement as to just what the science of thermodynamics was.

# 6

## Mysterious Fluids and Forces

As we have seen already, the new science of electricity appeared to promise a great deal to nineteenth-century people. The mysterious fluid seemed capable literally of performing wonders. Commentators and pundits waxed lyrical about the capacity of electricity to send signals almost instantaneously over vast distances. The electric telegraph was "a spirit like Ariel to carry our thought with the speed of thought to the uttermost ends of the earth."[1] Despite the gloomy prognostications of James Prescott Joule for one, that electricity could never economically replace steam, popular observers remained sanguine that electricity's unleashed power would provide the key to indefinite economic expansion and progress. As audiences flocked to exhibitions to witness more and more examples of electricity's wonders, electricity seemed to be the key that might unlock more of nature's secrets as well. Natural philosophers, electrical engineers, entrepreneurs, and showmen were all keen, therefore, to delve deeper into the strange fluid's mysteries. By the end of the nineteenth century, electricity had been joined by a plethora of other previously unheard-of fluids and forces. These new powers held out the promise of communication without wires, miracle cures for diseases, and even of ways of talking to the dead. In a world where the potentialities of science, technology, and human

[1] [A. Wynter], "The Electric Telegraph," *Quarterly Review*, 1854, 95: 119.

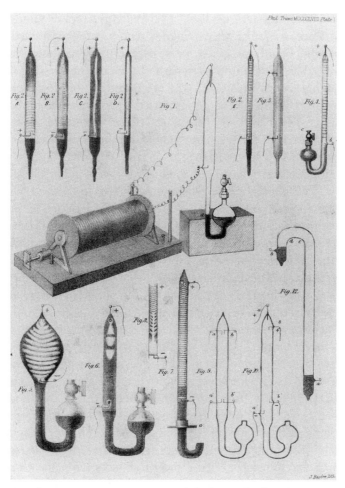

6.1 John Peter Gassiot's experiments with discharge tubes from the
*Philosophical Transactions*, 1858.

ingenuity appeared to some at least to be limitless, who was to say what
new powers science might find?

As early as the 1850s, men such as William Robert Grove and John
Peter Gassiot were working on the strange glowing effects produced by
passing electrical currents through partly evacuated tubes (figure 6.1).
Humphry Davy had shown back in the 1820s that the color of electrical
sparks varied according to the metal making up the poles. The glowing,
multicolored tubes seemed to be a variation on the same phenomenon.
By the 1870s more and more experimenters were working on trying

to explain these mysterious discharges. William Crookes thought they were evidence of a fourth state of matter, on top of the traditional triumvirate of solid, liquid, and gaseous states. Experimenters showed how the glows varied according to the strength of the currents passing through the vacuum tubes as well as the type of gas and its concentration in the tubes. Analysis of the discharges promised a new way of identifying the elements as well. The color of the glow seemed specific to the chemical elements present in the discharge tube. Experimenters discovered that magnets could be used to change the shape and even the direction of the strange discharges—cathode rays, as they were later called. Lecturers showed off photographs of the myriad shapes and arrangements the mysterious glows exhibited.

While many, if not most, practical telegraph and electrical engineers still tended to discuss electricity in terms of a fluid flowing down wires in just the same way that water flowed down a pipe, a new generation of theoretically inclined physicists—followers of James Clerk Maxwell's electromagnetic theories—took seriously the notion of electromagnetic fields. As far as they were concerned, most of the interesting things took place away from the wires that made up electrical circuits. They focused their attention on the field surrounding them. Self-taught mathematical physicist Oliver Heaviside developed his own sophisticated mathematical take on Maxwell's physics, showing how electrical energy moved through the ether. The experimenter and prolific public lecturer Oliver Lodge was convinced as well that ways could be found of detecting the propagation of electromagnetic waves through the ether. Both were pipped to the post in 1888 by the German physicist Heinrich Hertz, a student of Hermann von Helmholtz (himself one of the towering figures of nineteenth-century German physics) who announced to the world that he had found a way of propagating and detecting these long-sought-for electromagnetic waves. To Maxwellians in Britain it seemed to be the final proof that demonstrated the real existence of their master's hypothesized electromagnetic ether—after all, you could not have waves without a medium for them to move through.

Others were finding more novel ways of communicating at a distance. Late nineteenth-century Europe and America saw a groundswell of interest in spiritualist phenomena. More and more people claimed to be able to receive and transmit messages from the dead. Successful mediums (as they were called) became household names. While some scientists dismissed these men and women as charlatans, others argued that the phenomena they produced cried out for proper scientific investigation.

The Society for Psychical Research was established in 1882 for just that purpose. Maybe mediums' brains could somehow receive messages transmitted through the ether from beyond the grave. If so, then electrical theory and electrical experimentation were what was needed to find out what was going on. William Crookes was one of the most prominent of these psychical researchers. He put his expert knowledge of electricity to work in trying to find out whether mediums were lying when they claimed to be able to materialize ghosts that would walk around in Victorian drawing-room séances. It seemed to many that there was nothing preeminently outrageous about such claims. Electricity had already produced wonders—why should it not turn out to be a way of communicating with another world?

After the discoveries and breakthroughs of the 1880s, the 1890s seemed to many to be on the cusp of further and greater innovations. The search was on for more new forces and powers. Despite the expectations, Wilhelm Röntgen's discovery of some extraordinary rays in 1895 took the world by surprise. Röntgen seemed to have found a new kind of radiation that made it possible to see through solid objects. The mysterious X rays appeared to pass through solids in just the same way that rays of ordinary light passed through transparent materials such as glass. Within months of the discovery, X rays were already being applied in medicine. It took only a little longer for schoolboy jokes about the virtues of X-ray spectacles to start circulating. Within a few years, Marie and Pierre Curie in Paris had come up with another new kind of radiation. Radioactivity revealed a whole new world of energy, hidden away in apparently solid matter. The Curies' discovery opened up a new vista in physics too. It provided a window into the ultimate structure of matter and provided new clues as to how the building blocks of the universe might be held together.

The discovery and investigation of these mysterious fluids and forces during the second half of the nineteenth century played a pivotal role in the transformation of physics. As the new discipline became institutionalized in university laboratories across Europe and America, more and more budding practitioners turned to them as a source of experiments with which to make their reputations. They were turned into the raw material of a new culture of experimental research and theoretical speculation in physics. They demonstrated also how inseparable the new discipline still was from showmanship and exhibition. Doing physics still involved finding new ways of making the forces and powers of nature visible in as spectacular a way as possible. New technologies such as cathode ray tubes, radio transmitters, and X-ray photographs provided graphic evidence of humanity's hard-won power over nature and the possibilities for future progress. They were powerful arguments as well for those who wished to argue for the practical material benefits of training and research in physics for national economies. While these technologies formed the material basis for the consolidation of Maxwellian physics, however, they also planted the seeds of its downfall in the early decades of the twentieth century. An ambitious new generation of physicists could use these tools to refashion a new physics just as the previous generation had used them to bolster the old.

## Tubes That Glow in the Dark

Electrical enthusiasts throughout the early part of the nineteenth century were continually on the lookout for spectacular new experiments that could be used to make electricity visible. Particularly following Oersted's discovery of electromagnetism and the proliferation of demonstration devices that followed from Ampère, Barlow, Faraday, and Sturgeon, among others, a whole technology of display was developed with the aim of making it possible for audiences to see electricity in action. As we saw earlier, apparatus for showing the production of mechanical effects by means of electromagnetism, for demonstrating electrical shocks, and for exhibiting chemical effects such as the decomposition of water took pride of place in lecture theaters and exhibition halls as well as laboratories. Some of the most spectacular effects that electricity could produce, however, involved the electric spark. Striking multicolored scintillations could be produced between the poles of powerful galvanic batteries. Different colors could be produced by using different metals for the poles. Copper produced one color, platinum another, and so on. These powerful discharges had a utility beyond their impressive appearance. The powerful light they produced seemed a good candidate for commercial exploitation. As a result of all this, by the 1840s much experimental attention was being devoted to the physics of the electric spark and the circumstances surrounding its production.

William Robert Grove had been experimenting with electrical discharges from at least 1840. In particular, working with his newly invented nitric acid battery, he had been carrying out a series of experiments on the appearance and behavior of electrical sparks in different media, arguing that "the voltaic arc bears a similar relation to common flame, to that which electrolysis bears to ordinary chemical action."[2] By the 1850s, Grove was long departed from his professorship of experimental philosophy at the London Institution and the resources that position provided him. His experimental work was increasingly focused on working out the phenomena associated with electrical discharges under various conditions. In 1852 he published in the *Philosophical Transactions* of the Royal Society his observations that when the electrical discharge between two poles took place inside a tube evacuated of air, a diffuse glow was visible and that dark gaps or bands could also be seen at various points along the tube. He accounted for these bands in terms of the interrupted electrical discharges produced by the coil. Grove continued with this work for the following decade. In 1858 he described as well how he had found some interesting magnetic effects on the discharge. It seemed that by moving magnets around the tube in which the electrical discharge was taking place, he could make it change its shape and direction, just as Davy in the 1820s had found he could deflect an electric arc with a magnet.

Grove had come across the possibility of using magnets to manipulate the electric discharge in the work of the University of Bonn physicist Julius Plücker. Plücker had been struggling with his position in Bonn for more than a decade. A student there, he had originally been appointed a *Privatdozent* in mathematics and physics there in 1825, being promoted to extraordinary professor of mathematics in 1828 and then to ordinary professor in 1835. For most of the 1840s and early 1850s he was to all intents and purposes in charge of the physics cabinet at the university as well, despite efforts to persuade him to restrict his attention to his mathematical teaching. His discharge tube experiments in the late 1850s were in many ways the culmination of his experimental career. He observed that when a simple point was used as a negative electrode in his experiments, the glowing light between the electrodes was concentrated along the line of the magnetic force passing through that point as if the glow was following the line of the magnetism. He found, as well, that the walls of the vacuum tube itself glowed under the effect of the discharge and that holding magnets near the tube could alter the position of that glow.

John Peter Gassiot, Grove's friend and ally in ongoing battles to reform the Royal Society, was also involved in experimenting on electrical

[2]W. R. Grove, "On some Phenomena of the Voltaic Disruptive Discharge," *Philosophical Magazine*, 1840, 16: 482.

discharges during the 1850s. Gassiot, once a leading member and treasurer of the London Electrical Society, was the scion of a wealthy wine merchant family. He had the wherewithal to invest in experiment on a scale that few of his contemporaries could. The "electrical soirées" at his house in Clapham Common during the early 1840s are a good indication of his fascination with making electrical effects spectacularly visible. Like Grove, too, he was interested in the striae, or "stratifications," of the electrical discharge and the conditions under which they were produced. He could experiment on a large scale, noting that he "had the opportunity of experimenting with upwards of sixty of Geissler's vacua-tubes, in which many beautiful and novel results are produced; in some, for several seconds after the discharges had ceased, the tubes remained throughout their entire length highly phosphorescent."[3] The mercury air-pump, developed in 1855 by German glassblower Heinrich Geissler, had dramatically improved the production of evacuated tubes for such experiments. Gassiot took the view that the stratifications in the discharge "arose from pulsations or impulses of a force acting on highly attenuated matter."[4]

Discharge experiments like these rapidly became established as part of the technology of display of electrical performers of various kinds. In British eyes at least, the new technology remained intimately associated with the names of Grove and Gassiot. Cromwell Varley, in a communication to the Royal Society on electrical discharges in rarified media in 1871, commenced with an effusive acknowledgment of "the labours of Mr. Gassiot" and an apology "lest he should appear to be attempting to appropriate the glory which so justly belongs to that gentleman and to Professor Grove."[5] Varley, a leading telegraph engineer, was particularly interested in using photography as a means of capturing the appearance of the glowing tubes. Photographic technology could capture images that were beyond the power of the naked eye. "The light was so feeble that, though the experiment was conducted in a perfectly dark room, we were sometimes unaware whether the current was passing or not. An exposure of thirty minutes' duration left, as will be seen, a very good photographic record of what was taking place." Varley's description of the way in which the discharge glow developed with increased current

---

[3] J. P. Gassiot, "On the Stratification in Electrical Discharges," *Philosophical Transactions*, 1859, **149**: 137.

[4] Ibid., 156.

[5] C. Varley, "Some Experiments on the Discharge of Electricity through Rarified Media and the Atmosphere," *Proceedings of the Royal Society*, 1871, **19**: 236.

was striking: "a tongue of light projected from the positive pole towards the negative, the latter being still completely obscure. The light around the positive pole was to all our eyes white, while the projecting flame was a bright brick-red."[6]

The most prolific and energetic researcher into discharge phenomena during this period was William Crookes. Crookes, the son of a well-off businessman and a successful entrepreneur in his own right, was an enthusiastic experimenter. He had studied at the Royal College of Chemistry and been impressed by Michael Faraday's lectures at the Royal Institution. Early in his career, he attracted the attention and patronage of the natural philosopher and telegraph entrepreneur Charles Wheatstone, who steered him in the direction of his first Royal Society grant. During the 1860s, Crookes was particularly concerned with elucidating the behavior of a curious piece of apparatus he called a radiometer. Puzzled by the apparent gain in weight of substances weighed in a vacuum when they were heated, Crookes built an instrument in which a delicately balanced vane, enclosed in a flask from which the air had been removed, could be made to rotate under the influence of heat or light. Crookes first argued that the movement was caused by pressure exerted by the radiation itself. He soon changed his mind however, arguing instead that residual air molecules in the flask caused the movement. He raised the question, however, of whether the substance remaining in the flask "should not be considered to have got beyond the gaseous state, and to have assumed a fourth state of matter, in which its properties are so far removed from those of a gas as this is from a liquid."[7] Looking for further examples of his newly proposed "fourth state of matter" Crookes lighted upon those curious glowing electrical discharges.

His interest was initially captured not so much by the glowing discharges themselves, but by the dark space that observers agreed could be seen around the negative pole in these kinds of experiments. Crookes interpreted this dark space as further evidence of his fourth state of matter. He argued that it was the result of "molecular pressure" of the same kind that caused movement in his radiometer. "This dark space is found to increase and diminish as the vacuum is varied in the same way that the ideal layer of molecular pressure in the radiometer increases and diminishes. As the one is perceived by the mind's eye to get greater, so the

[6]Ibid., 238.

[7]W. Crookes, "Experimental Contributions to the Theory of the Radiometer," *Chemical News,* 1876, 34: 277.

other is seen by the bodily eye to increase in size."[8] In Crookes's view, the difference between ordinary gases and his fourth state of matter lay in the hugely increased mean free path of the gas molecules—that is to say the average distance the molecules traveled before colliding with each other. This mean free path was much longer in the fourth state, and according to Crookes the dark space was a measure of it. "The thickness of the dark space is the measure of the length of the path between successive collisions of the molecules. The extra velocity with which the molecules rebound from the excited negative pole keeps back the more slowly-moving molecules which are advancing toward the pole. The conflict occurs at the boundary of the dark space, where the luminous margin bears witness to the energy of the discharge."[9]

According to Crookes, as the vacuum in the discharge tube was increased, so did the mean free path of the molecules. This explained the way in which the glass of the tube itself appeared to phosphoresce at a very high vacuum. When the mean free path of the negatively charged molecules streaming across the tube reached the same length as the size of the tube, the molecules collided with the sides of the tube as well as with each other, causing the glass to glow. When a cross was placed between the negative electrode and the wall of the tube, a shadow was formed on the glass where the stream of molecules was prevented from hitting it. Crookes even found that when the cross was removed, the area that had been in its shadow now glowed more brightly. "Here, therefore is another important property of Radiant Matter. It is projected with great velocity from the negative pole and not only strikes the glass in such a way as to cause it to vibrate and become temporarily luminous while the discharge is going on, but the molecules hammer away with sufficient energy to produce a permanent impression upon the glass."[10] One of Crookes's most striking illustrations of the power of these streams of radiant matter is also a graphic illustration of the exhibitionist tendency of his experiments. In this experiment, a little glass railway, carrying a tiny locomotive with a paddle wheel was placed between aluminum poles in an evacuated tube. The streams of radiant matter flowing between the

[8]W. Crookes, "Molecular Physics in High Vacua," *Proceedings of the Royal Institution*, 1882, 9: 140.

[9]W. Crookes, "On the Illumination of Lines of Molecular Pressure," *Philosophical Transactions*, 1879, 170: 135.

[10]W. Crookes, "On Radiant Matter," *Chemical News*, 1879, 40: 106.

poles would strike the paddle wheel, causing it to rotate and propel the little locomotive along its miniature track.

Few of his contemporaries found Crookes's claims concerning the fourth state of matter convincing. In particular the length of mean free path for molecules required by his notions of the fourth state were glaringly at variance with those posited by the kinetic theory of gases. For a physicists' culture that increasingly lauded high mathematical theory, Crookes's experiments, however spectacular, could never ultimately compete with the theory-laden pronouncements of a Maxwell. Even a mathematical physicist such as George Gabriel Stokes remained an admirer of Crookes's work, however. "For enlarging our conceptions of the ultimate workings of matter, I know of nothing like what Crookes has been doing for some years," he marveled to a friend. "I wish you could see some of the work in his laboratory."[11] Enthusiasm for discharge tube (or cathode ray tube) experiments continued. Plücker's student Wilhelm Hittorf had been working on them throughout the 1870s, having published work on magnetic deflections and the production of shadows as early as 1868. In Britain, Warren de la Rue and Hugo Müller carried out detailed experiments aimed at elucidating the exact circumstances under which the discharges were produced to best effect and producing some stunning illustrations along the way. They were insistent that far from being some manifestation of a fourth state of matter, the "electric arc and the stratified discharge are modifications of the same phenomenon."[12]

Experiments with cathode ray tubes were at the cutting edge of experimental physics for much of the second half of the nineteenth century. Their performers were convinced that understanding those mysterious glowing tubes would provide the key that could crack open the secrets of matter, even if they disagreed over just what those secrets might be. Everyone agreed that powerful forces were at play inside those tubes that could tear apart the bonds that usually held matter together. Exhibiting those powerful forces was a matter of some concern to these experimenters as well. William Crookes certainly kept a weather eye on the show-off potential of his experimental apparatus even as he and his assistants were working away in the laboratory. Crookes, as a self-made man who plied his scientific trade outside the walls of academe, was

[11] Quoted in R. deKosky, "William Crookes and the Fourth State of Matter," 58.

[12] W. de la Rue and H. Müller, "Experimental Researches on the Electric Discharge with the Chloride of Silver Battery," *Philosophical Transactions*, 1880,171: 109.

particularly alert to such possibilities. The strange, flickering glows inside the discharge tubes were a potent image of the mysteries of nature that modern physics promised to lay bare. They pushed at the boundaries of mundane reality and posed the question of what else was out there waiting to be uncovered. They emphasized the newfound powers of their creators as well—and their claims to authority in the modern world.

## Waves in the Ether

Developments like Crookes's and his fellow-experimenters' work on cathode rays made it clear to their contemporaries just how much more lay out there, waiting to be discovered. The search was now well and truly on for new and mysterious manifestations of nature's forces. The discovery of new phenomena like these could provide forceful demonstrations of the power of modern physics and its practitioners. They were graphic illustrations of the ways in which physicists could impose their mastery over nature. They promised new utilitarian advances as well. Telegraphy had already demonstrated how electricity could revolutionize the world and by the 1870s other electrical technologies were making their mark too. Who was to say what equally unprecedented advances might be made on the backs of other novel discoveries? These new findings provided powerful backing for new theoretical generalizations as well. For many ambitious young physicists, in Britain in particular, they provided the final proof of the existence of an electromagnetic ether—that the ether was the medium through which electromagnetic as well as simply optical phenomena manifested. Looking back from the vantage point of the 1890s, Oliver Lodge was adamant as to what the past two decades' experiments demonstrated: "Persons who are occupied with other branches of science, philosophy, or with literature, and who have not kept quite abreast of physical science, may possibly be surprised to see the intimate way in which the ether is now spoken of by physicists, and the assuredness with which it is experimented on. They may be inclined to imagine it is still a hypothetical medium whose existence is a matter of opinion. Such is not the case."[13]

The publication of James Clerk Maxwell's *Treatise on Electricity and Magnetism* in 1873, with its rich theoretical synthesis, laid the groundwork for a new generation of his followers, committed to his view of the ether as the anchorage for electromagnetic energy. This perspective,

[13]O. Lodge, *Modern Views of Electricity* (London, 1892), viii.

which drew the physicist's attention away from the coils and wires of electrical apparatus and towards the apparently empty space around them, was at gross variance with the approach of the hands-on electrical engineers who dealt with such apparatus every day as they maintained the nation's telegraph lines. Their view of electricity was robustly simple. It was just like water running down a pipe. Few of them had much time for Maxwell and his newfangled ideas, which seemed to bear little relevance to their everyday experiences. Maxwell's followers, on the other hand, saw this ongoing battle of practice versus theory as an ideal opportunity to press their own claims to expertise over and above those of the practical men. If they could show that they understood better than the engineers themselves what was going on in the telegraph network, then they would have shown that theory really mattered. It would be a major boost for the cultural authority of physics.

The major protagonists in this war between physicists and practical men were William Henry Preece—who ran Britain's nationalized telegraph network through the Post Office—for the practicals, and the two Olivers, Heaviside and Lodge, for the theoreticals (figure 6.2). Heaviside, a self-trained mathematician of considerable brilliance, had been working on his own reformulation of Maxwell's theory, producing along the way the four Maxwell's equations that are central to modern understandings of Maxwell's work. His work drew attention in particular to the problems of self-induction in rapidly oscillating currents—self-induction being the tendency first noted by Faraday for an electric current to oppose changes in its own strength. According to Maxwell's (and Heaviside's) theory, the faster the oscillations, the less like a fluid in a pipe electricity became. In very rapidly alternating currents, the electricity ran almost entirely along the surface of the wire rather than along the interior. This meant that under such circumstances, self-induction rather than the resistance of the wires became the major factor in designing cables. Preece dismissed this as nonsense. It was no more than "a bug-a-boo." Lodge encountered self-induction as well in his work on lightning. He argued in lectures before the Society of Arts that self-induction mattered far more than resistance in the design of lightning conductors and that traditional conductors designed for low resistance were useless for handling sudden lightning bolts. Preece, a member of the 1882 Lightning Rod Conference whose conclusions Lodge was attacking, dismissed his conclusions as absurd.

As far as the Maxwellians were concerned, the key to establishing Maxwell's theory and demolishing the practicals' pretensions to down-to-earth knowledge was to find a sure way of showing that electromagnetic

6.2 The satirical magazine *Punch's* view of the debate between scientific and practical electricians in 1888. The triumphant practical man William Henry Preece is walking all over the recumbent Maxwellian, Oliver Lodge.

energy really did travel through the ether. In the early 1880s, the Dublin professor George Francis FitzGerald had worked out in some detail the theory of the matter and had calculated the amount of energy that would be given off. The trick was to find some way of detecting those mysterious vibrations in the ether. Oliver Lodge, by 1881 professor of physics at Liverpool, was determined to solve the problem. Lodge had been intrigued by the ether for years: "At an early age I decided that my main business was with the imponderables—as they were then called—the things that worked secretly and have to be apprehended mentally. So it was that electricity and magnetism became the branch of physics which most fascinated me."[14] Lodge was no Cambridge-trained mathematician, though. His route to understanding "things that worked secretly" would be through experimentation. He modeled his Liverpool laboratory on the lines recommended by the illustrious Sir William Thomson and traveled to Germany to pick up the best equipment available.

One of Lodge's projects was to investigate lightning at the behest of the Royal Society of Arts. It was a practical project aimed at improving the design of lightning rods. As a practical electrical experimenter in the tradition of William Sturgeon, Lodge set about finding a laboratory model for the way lightning worked and struck upon the Leyden jar. Lightning was like the discharge from a Leyden jar magnified a thousandfold. This had an important consequence. As all electricians knew, Leyden jar discharges oscillated rapidly. This meant that lightning did so as well. Far from being a single discharge of electricity from heaven to Earth, a stroke of lightning was a rapid succession of discharges in both directions. The trick to understanding lightning, for Lodge, was to carefully investigate the characteristics of Leyden jar discharges in his laboratory. Lodge wanted to show that what mattered in such situations was self-induction rather than resistance. If he showed in his Leyden jar experiments that the discharges would follow a path of high resistance and low self-induction, rather than one of low resistance and high self-induction, he would have shown that was how lightning behaved as well. It was his announcement of the results of these experiments that so aroused the ire of William Henry Preece at the Post Office, incensed by the interference of a mere physicist in the affairs of hardheaded practical men.

Lodge soon realized, however, that more was at stake in his experiments than lightning rods. His experiments with Leyden jars convinced him that he was well on his way to finding ways of producing and—more importantly—detecting those elusive oscillations in the ether that FitzGerald's work predicted. The electrical currents that surged back and forth along the wires of his experimental apparatus in imitation of the lightning could be made, by proper arrangement, to stabilize into static, standing waves that could be measured. This was his next task. In the run-up to the British Association for the Advancement of Science's meeting

---

[14] O. Lodge, *Past Years* (London: Stodder & Haughton, 1931), 111.

at Bath in 1888, Lodge worked hard to produce a spectacular experimental demonstration to convince the crowds of the existence of electromagnetic waves. "A long wire about one-eighth or one-sixteenth of an inch in diameter was stretched round the theatre at South Kensington, several times round," he recalled in his autobiography. "On now sending a series of oscillating pulses along that wire, a glow was to be seen on the wire in the dark at the ventral segments of each pulse, while it was dark at the nodes, exactly in accord with Melde's experiments of a string attached to a tuning-fork . . . The point of the experiment was not another demonstration of the oscillations, but a proof that true waves ran along the wires, being thereby guided and prevented from spreading into space, and by reflexion were converted into stationary waves which showed themselves by nodes and loops."[15] He had found what he was looking for; he had vindicated himself and his fellow Maxwellians against the recalcitrant practical men. He was too late, however. Before the Bath meeting the young German physicist Heinrich Hertz announced to the world his detection of electromagnetic waves in space (figure 6.3).

Hertz was in many ways the golden boy of German physics. He was one of Helmholtz's Berlin products, having turned to physics after an early flirtation with engineering. He had been trained, therefore, by one of the grand masters of German physics, one moreover who had made his reputation as an experimentalist as much as he had as a theorist. Working with Helmholtz in his laboratory, Hertz slowly built up a reputation for himself during the late 1870s and early 1880s as a diligent and skilled experimenter. His early interest in engineering gave him a distinct edge over many of his German contemporaries. The view of electromagnetism that Hertz acquired from his teacher was very different from that held by his Maxwellian contemporaries in Britain. Helmholtz had little time for Maxwell's electromagnetic ether filling all space. Action took place at a distance rather than directly through an intervening medium. Hertz first made a name for himself as an independent experimenter with an experiment designed to confound the theories of Helmholtz's rival in the field of German electricity, Wilhelm Weber. By the mid-1880s, when Hertz was his own man first at Kiel and later from 1885 at Karlsruhe, he was still working on problems largely defined by Helmholtz's perspective on physics and aimed at bolstering his mentor's theories against local German adversaries.

[15]Ibid., 183–84.

6.3 Some of the apparatus that Heinrich Hertz used to demonstrate the existence of electromagnetic waves in the ether.

At Karlsruhe, Hertz initiated a series of experiments into the properties of electrical discharges. He still saw himself as working on problems initiated by Helmholtz—and in particular on a prize question proposed by Helmholtz in Berlin as early as 1879 on the electrical effects of dielectrics (substances like air or glass that do not, under normal circumstances, conduct electricity) on neighboring conductors. In his Karlsruhe laboratory he became increasingly intrigued by the properties of some of the demonstration apparatus. In particular, he was interested in the sparking effects of coils arranged so that they were adjacent to each other. Hertz worried away at the problem, trying to find better ways of experimentally reproducing the phenomena. After spending some time trying (and failing) to get rid of some irritating "side sparks" that detracted from what he regarded as the main phenomenon, he gave up in disgust and switched his attention to the side sparks themselves. He found he could use these side sparks as indicators of surges of current across nearby spark gaps (in other words, dielectrics). They could be used to probe for interesting electrical effects. Over the next few years, he gradually built up a repertoire of what seemed to be wavelike effects by manipulating his apparatus. He could demonstrate nodes in standing waves. He could detect reflections and even interference patterns just as if the mysterious waves were rays of light. Finally, in 1888 he revealed his experiments to the world.

Looking back at his experimental endeavors of the past few years from the vantage point of a successful conclusion, Hertz now placed his work firmly in the context of Maxwellian physics. His electric waves were a vindication of Maxwell. While the experiments and the phenomena they produced stood independently of any theory, their significance largely lay in the way they sorted out the wheat from the chaff in electromagnetic theory. "Since the year 1861 science has been in possession of a theory which Maxwell constructed upon Faraday's views, and which we therefore call the Faraday-Maxwell theory. This theory affirms the possibility of the class of phenomena here discovered just as positively as the remaining electrical theories are compelled to deny it . . . as long as Maxwell's theory depended solely upon the probability of its results, and not on the certainty of its hypotheses, it could not completely displace the theories which were opposed to it. The fundamental hypotheses of Maxwell's theory contradicted the usual views, and did not rest upon the evidence of decisive experiments. In this connection we can best characterise the object and the result of our experiment by saying: The object of these experiments was to test the fundamental hypothesis of the Faraday-Maxwell theory, and the result of the experiment is to confirm the fundamental hypothesis of the theory."[16] His electric waves were to be interpreted as the jewel in the crown of Maxwellian physics.

Lodge in Liverpool responded to Hertz's triumph with commendable humility, his chagrin tempered presumably by the knowledge that these Hertzian waves provided some very heavy ammunition for his battle with the practical men. He was unequivocal in his interpretation of their significance: "In 1865, Maxwell stated his theory of light. Before the close of 1888 it is utterly and completely verified. Its full development is only a question of time, and labour, and skill. The whole domain of Optics is now annexed to Electricity, which has thus become an imperial science."[17] Lodge was soon out demonstrating electric waves to excited audiences, with sometimes spectacular results. As he recalled, "I exhibited many

[16]H. Hertz, *Electric Waves* (London: Macmillan, 1893), 19–20.
[17]O. Lodge, *Modern Views of Electricity* (London, 1892), 336.

of the Leyden Jar experiments both to the Royal Institution and the Society of Telegraph Engineers, in a lecture on 'The Discharge of a Leyden Jar,' where were shown many striking experiments. The walls of the lecture-theatre, which were metallically coated, flashed and sparked, in sympathy with the waves which were being emitted by the oscillations on the lecture-table—an incident which must be remembered by many of those present. This was a novel result, surprising to myself also, and I hailed it as an illustration or demonstration of the Hertz waves."[18] When Hertz died at the early age of thirty-six in 1894, Lodge paid generous tribute to his memory in a lecture at the Royal Institution.

Other Maxwellians were quick to jump onto the Hertzian bandwagon. William Crookes concurred with Lodge in his assessment of what had happened. "Whether vibrations of the ether, longer than those which affect us as light, may not constantly be at work around us, we have, until lately, never seriously enquired. But the researches of Lodge in England and of Hertz in Germany give us an almost infinite range of ethereal vibrations or electric rays, from wave-lengths of thousands of miles down to a few feet." The business-minded Crookes was not slow to grasp the commercial possibilities either. "Rays of light will not pierce through a wall, nor, as we know only too well, through a London fog. But the electrical vibrations of a yard or more in wave-length of which I have spoken will easily pierce such mediums, which to them will be transparent. Here, then, is revealed the bewildering possibility of telegraphy without wires, posts, cables, or any of our present costly appliances." This was, he insisted "no mere dream of a visionary philosopher. All the requisites needed to bring it within the grasp of daily life are well within the possibilities of discovery."[19] Nor was telegraphy the only possibility. Crookes speculated that the "ideal way of lighting a room would be by creating in it a powerful, rapidly-alternating electrostatic field, in which a vacuum tube could be moved and put anywhere, and lighted without being metallically connected to anything."[20] Crookes was quite right. By the 1890s Guglielmo Marconi was already experimenting with wireless telegraphy under the aegis of the British Post Office. Nikola Tesla was experimenting to realize the dream of electric light without wires. It all went to confirm Oliver Lodge's robust claim that one could no more deny the existence of the ether than one could deny the existence of matter.

[18] O. Lodge, *Past Years* (London: Stodder & Haughton, 1931), 185.
[19] W. Crookes, "Some Possibilities of Electricity," *Fortnightly Review*, 1892, 51: 175.
[20] Ibid., 177.

Not everyone subscribed to this imperialist reading of Hertz's experiments, however. When George Francis FitzGerald asserted at the Bath meeting of the British Association that the existence of electric waves was the final proof of Maxwell's theory, an opponent from the practical men's camp retorted that so far as he could see "the Hertz experiments prove nothing."[21] From the practicals' perspective, the Maxwellians' efforts to use their expertise as physicists to adjudicate over matters of practical, hands-on electricity were impertinent attempts by young upstarts to foist their book learning on men who had imbibed their craft through hard-won experience. This kind of debate underlines the point that, as in many other cases, what was at stake here was authority. Physics and physicists had to find a cultural role for themselves if the new discipline was to be ultimately successful. Hertz's waves were from this perspective a gift, albeit a hard-fought-for one. They demonstrated graphically the physicist's power to manipulate nature, to make things happen as if by magic. For large sections of the public they raised the possibility as well that discoveries of more mysterious forces in the ether were yet to be made and that other kinds of transmitters and receivers were waiting to be found.

### The Other World

Many late nineteenth-century commentators clearly felt that new discoveries like cathode rays and electric waves were only the tip of the iceberg. A whole order of forces was waiting to be discovered in the ether. Crucially, many argued that physics was on the verge of delivering answers to fundamental questions of life and death. The Victorian period was in many ways the heyday of alternative scientific practices. Mesmerism and table turning flourished in the early part of the period. There was intense interest in spiritualism around the end of the century. For many (though by no means all) of their practitioners, these practices were not alternatives to science—they were scientific themselves. The phenomena they produced could be studied and manipulated in just the same way as physicists could study and manipulate electricity or the ether. Significant numbers of physicists in Europe and America concurred with this assessment. A far larger number disagreed vociferously. Practitioners of mesmerism or spiritualism were either frauds or dupes and men of science who took their practices seriously were themselves deceived.

---

[21] Quoted in B. Hunt, "Practice vs. Theory," 351.

Nevertheless, such phenomena were the subject of serious inquiry. To some physicists at least, they seemed ripe for incorporation into the new physics of the ether.

An early example of such an effort was the work of the German natural philosopher Karl von Reichenbach. His announcement in 1845, in the pages of the *Annalen der Chimie und Pharmacie*, edited by the illustrious Justus Liebig, that he had discovered a new kind of force was greeted with considerable interest. This odic force, as he called it, seemed to emanate from the poles of magnets and from living beings. It could be sensed only by particularly sensitive female subjects—typically, hysterical or nervous women. Such subjects could actually see the new force as a flame emanating from the poles of magnets or as an aura surrounding living matter. Reichenbach was emphatic that what he had discovered was a physical force just like electricity or magnetism, albeit one that seemed to have a particularly close affinity with the processes of life. He and others speculated that it might be correlated with other physical forces and therefore fitted in to the grand new universal systems of interrelated forces being offered by natural philosophers like William Robert Grove. Odic force was a subject of intermittent experimental research for much of the century. Its apparent close relationship to living matter made it a particularly intriguing topic of inquiry. A proper understanding of odic force's relationship to other forces might deliver the secret of life itself.

Many enthusiasts for odism had interests in another flourishing mid-century practice as well—mesmerism. Mesmerism, or animal magnetism, already had a long history by the Victorian period. The practice was developed in the late eighteenth century by the Viennese physician Anton Mesmer. He argued that living matter contained an innate animal magnetism that he and similarly adept practitioners could manipulate in various ways to produce trancelike states in their subjects. A number of early Victorian natural philosophers, like the radical doctor John Elliotson and the popular writer and lecturer Dionysius Lardner, professor of mechanical philosophy at University College, were convinced of mesmerism's reality. To Lardner it was clear that the mysterious agent behind the phenomena "is material, is propagated through space in straight lines; that various corporeal substances are pervious by it with different degrees of facility, and according to laws which still remain to be investigated; that it is reflected from the surfaces of bodies, according to definite laws, probably identical with or analogous to those which govern the reflection of other physical principles, such as light and heat; that it has a specific action on the nervous systems of animated beings, so as to produce in them perception and sensation, and to excite various mental emotions."[22] It was just like any other force of nature.

Even many of the mesmerists' most committed opponents found it difficult to deny the reality of the phenomena that could apparently be produced by animal magnetism. Their strategy instead was to deny the existence of the magnetic fluid that, according to the mesmerists, caused the phenomena. In effect, they tried to transform mesmerism from a practice that might legitimately be studied by physicists into one that was best dealt with by another rising new discipline—psychology. According to the physiologist William Benjamin Carpenter, for example, the mesmeric trance was simply the result of pathological processes in the brain rather than the outcome of some mysterious force: "ideas which take possession of the mind, and from which it cannot free itself, may excite respondent movements; and this may happen also when the force of the idea is morbidly exaggerated, and the will is not suspended but merely weakened, as in many forms of insanity."[23] People in mesmeric trances were simply people who had allowed themselves to lose control over their own minds and bodies. Not everyone was convinced, however. Even as interest in mesmerism waned significantly during the second half of the century, many experimenters remained convinced that there was some hitherto undiscovered connection between mental and nervous processes and the operations of the ether.

Many attempts to make sense of the human mind and body in terms of electrical processes had been made by the final decades of the nineteenth century. The telegraph, in particular, was an increasingly common metaphor for the nervous system from midcentury onwards. The nerves were characterized as telegraph wires carrying messages to and fro between the brain and the body's peripheries. Electrophysiologists were thought to have "tapped the wires of the living telegraphic system."[24] The brain itself from this perspective was a mass of electrical circuits. "Could we picture to ourselves the changes in the brain when its higher centres are in a state of molecular disturbance, as when one is thinking rapidly . . . could we, in such circumstances of mental turmoil, examine the phenomena of the brain, we could, in all probability, obtain evidence

[22]Quoted in A. Winter, *Mesmerized*, 55.

[23]W. B. Carpenter, *Human Physiology* (London, 1853), 672.

[24]J. G. McKendrick, "Human Electricity," *Fortnightly Review*, 1892, 51: 638.

of rapid changes of potential, and of currents flashing in a thousand directions, pursuing paths the intricacies of which are many times greater than if all the telegraphic and telephonic wires of London were concentrated in one vast exchange."[25] As ether physics gained ground over its opponents towards the end of the century it seemed to offer a new way of looking at the human mind and body as well.

Oliver Lodge was one of the first to suggest that the latest electric wave technology could be used to model the body. The eyes, he suggested, acted just like coherers—a kind of detector for electric waves. Others took up the suggestion enthusiastically. The medical electrician William Hedley argued that Lodge's work showed how human beings interacted with the ether. "The conductor, whether it be a wire or a living body, only *guides* the energy and concentrates it for useful work. Applying this to that very imperfect conductor, the human body, it is evident that the latter may be regarded as an appliance capable of utilising in a variety of ways energy transmitted by the ether."[26] The sense organs were a set of receivers "syntonised for the reception of similarly vibrating etherial impulses radiating from some given source."[27] Others took matters further. William Crookes speculated that "other sentient beings have organs of sense which do not respond to some or any of the rays to which our eyes are sensitive, but are able to appreciate other vibrations to which we are blind . . . Imagine, for instance, what idea we should form of surrounding objects were we endowed with eyes not sensitive to the ordinary rays of light but sensitive to the vibrations concerned in electric and magnetic phenomena."[28] Furthermore, he argued, "In some part of the human brain may lurk an organ capable of transmitting and receiving other electrical rays of wave-lengths hitherto undetected by instrumental means. These may be instrumental in transmitting thought from one brain to another."[29]

These kinds of speculations seemed to some, at least, to offer possible answers to puzzles like Reichenbach's odic forces. Maybe the high-strung women who claimed they could see flames emanating from the poles of a magnet were somehow attuned to be able to see magnetic forces. Jean Charcot in Paris carried out experiments to assess the effect of magnets

[25] Ibid., 639.

[26] W. S. Hedley, "Apologia pro Electricitate Suâ," *Lancet*, 1895, 1: 1103.

[27] Ibid.

[28] W. Crookes, "Some Possibilities of Electricity," *Fortnightly Review*, 1892, 51: 176.

[29] Ibid.

on sensitive, hysterical women. William Henry Stone, medical electrician, member of the Society for Psychical Research and physician at St. Thomas's Hospital in London, carried out similar experiments in an effort to check Charcot's conclusions. He found some evidence that placing his patient's head between the poles of a powerful electromagnet alleviated her headaches. He felt bound to add though that having tried the experiment on himself, he could notice no definite result. William Barrett, professor of experimental physics at Dublin's Royal College of Science, also carried out experiments under the aegis of the Society for Psychical Research to investigate Reichenbach's odic forces. He concluded that there was a strong case in favor of the existence of some peculiar and unexplained luminosity around the magnetic poles, which could only be seen by certain individuals.

For William Crookes, however, ether physics offered something far more revolutionary than a way of explaining mysterious magnetic auras. It offered a way of communicating with the dead. Spiritualism, as the practice came to be called, was increasingly popular towards the end of the nineteenth century and the beginning of the twentieth. Spiritualism offered a system of practices whereby people could commune with the spirits of the dead. Mediums offered themselves as conduits through which spirit guides could pass messages from the dead to the living, offering condolences, imparting information, and describing their existence in the spirit world. While many spiritualists quite explicitly cast their practices in direct opposition to what they regarded as the excessively materialist stance of modern science, some physicists regarded spiritualist phenomena as being ripe for physical explanation in terms of the ether. Spiritualism was yet another example of the ether's mysterious powers waiting to be unlocked. This was the kind of thing Crookes had in mind with his talk of sentient beings with sense organs responsive to different wavelengths of electromagnetic rays and of humans with organs in their brains that could receive and transmit thought. He and similarly minded physicists were deeply involved in investigating spiritualist phenomena. To many of his fellow physicists it seemed a very dangerous road to follow. He was in danger of becoming the tool of charlatans (if not becoming a charlatan himself).

Crookes and his fellow electrician, the telegraph engineer Cromwell Varley, had been involved in investigating spiritualist phenomena since the 1860s. Electricians, dealing as they did with a mysterious force that seemed to act at a distance through some medium or other, seemed well-qualified to investigate such matters. Some spiritualists agreed with this assessment. The American spiritualist Emma Hardinge argued that her practices made use of "the self-same forces of the telegraph, worked by vital instead of mineral electricity."[30] Varley concurred, going so far as to invoke spiritualism in an attempt to explain the mysterious workings of the telegraph to a skeptical, working-class East End London audience. Making spiritualism scientific however, required the imposition of laboratory standards onto the séance. "Each circle should be under the management of a clever man and each should carry on a continuous and exhaustive examination of the groundwork of the subject. Once establish a clue to the relations existing between the physical forces known to us and those forces by which the spirits are sometimes able to call into play the power by which they produce physical phenomena—once establish this clue there will be no lack of investigators, and the whole subject will assume a rational and intelligible shape to the outside world."[31]

Crookes and Varley entered into their spiritualist investigations by bringing all the impedimenta and paraphernalia of modern science to bear on the problem. Crookes imported the latest and most advanced laboratory precision measurement apparatus into the séance in an effort to measure the psychic force exerted by the medium Daniel Dunglas Home in the early 1870s. Convinced that not only mediums but all human beings could manifest such forces to some degree, he searched hard for a way of reliably detecting such minuscule manifestations. It was no coincidence that these spiritualist researches took place at the same time as his investigations of the equally intangible forces manifested inside the radiometer. With Varley, Crookes devised an electrical means to detect imposture in cases of spirit manifestation—instances in which spirits seemed to become physically tangible and visible during the course of a séance. Photographs demonstrated the spirit's presence while the electric circuit showed that the medium—typically secluded in another room—was not moving during the course of the manifestation. According to Crookes and Varley, their experiments with Florence Cook and the ghostly Katie King provided tangible evidence that whatever else might be going on, Florence and Katie were not the same material person (figure 6.4). As Varley insisted, even an experienced electrician would simply find it impossible to escape the circuit without springing the trap.

---

[30] E. Hardinge, "What is Spiritualism?" *Human Nature*, 1867, 1: 568–69. My thanks to Richard Noakes for this reference.

[31] Quoted in R. Noakes, "Telegraphy Is an Occult Art," 445.

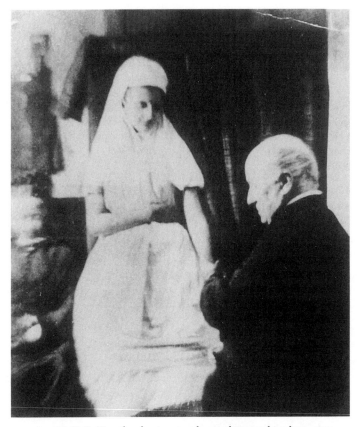

6.4 Katie King the ghost captured on a photographic plate.

Oliver Lodge—almost the high priest of ether physics—was also increasingly interested in the possibilities of psychic forces manifesting themselves through the ether. He certainly concurred with Crookes's assessment that there was no reason to doubt that the ether might interact, or provide a vehicle for interaction, with living matter and mind in ways hitherto unimagined. Lodge had started experimenting seriously on such matters a few years after taking up his professorship in Liverpool in 1881. His experiments followed the spiritualist Irving Bishop's performances in the city in 1883, in which incidents of "thought transferences" were claimed to take place on stage. Asked to carry out some experiments on the phenomenon, he did so, being sufficiently impressed by the results to write them up for *Nature*. Somewhat to his surprise, the editor published them. Lodge was convinced that "thought-transference or ... 'telepathy'

from one person to another was a reality."[32] Looking back at the progress
of physics in his autobiography, Lodge declared his conviction that "the
scheme of physics will be enlarged so as to embrace the behaviour of
living organisms, under the influence of life and mind. Biology and psychology are not alien sciences . . . they belong to the physical universe,
and their mode of action ought to be capable of being formulated in
terms of an enlarged physics of the future, in which the ether will take a
predominant part."[33]

Many nineteenth-century physicists regarded the involvement of
some of their fellows in matters psychic as being distasteful at best and a
surrender to charlatanry at worst. Dabbling in mesmerism, odic forces,
or spiritualism was a betrayal of everything the young science of physics
stood for. Similarly, many spiritualists regarded the efforts of Crookes,
Lodge, and Varley to trespass on their territory as an impertinent irrele-
vance. Spiritualism was for them in direct opposition to science and had
no need for its support. Spiritualism from this perspective was an antidote to the oppressive materialism of scientific culture. Even spiritualists
sympathetic to physicists' aims were unhappy at their insistence on im-
posing laboratory protocols and disciplines on the culture of the séance.
For these physicists and their supporters, however, there was nothing
unusual about their interest—and nothing inimical to physics either. On
the contrary, they regarded their discipline's history as a vindication of
their stance. The progress of physics throughout the century was after
all punctuated by the discovery of new forces. There was in their view
nothing to distinguish psychic forces from the list of those already discovered. Far from it—they offered a rich new field of inquiry and another
opportunity to wield physics' newfound cultural authority.

## Radioactivity

By the last decade of the nineteenth century, Maxwellian physics, as Oliver
Lodge had indicated, appeared supremely triumphant. The discovery of
electric waves and the moves that were soon being made to transform
wireless telegraphy into a viable commercial technology placed the reality
of the electromagnetic ether beyond reasonable doubt. The ether was real
because it could be manipulated—it could be made to do things. It was
the ultimate vindication of the new physics' claim to cultural authority

---

[32] O. Lodge, *Past Years* (London: Stodder & Haughton, 1931), 275.
[33] Ibid., 350–51.

in the modern age. Pundits speculated as to what new properties of this universal medium might next be discovered and placed at the service of humanity through the powers of physics. Writers in the new genre of science fiction penned stories in which the citizens of future utopias lived lives of unparalleled luxury and leisure, all fueled with energy harvested from the ether. Other more pessimistic authors warned of a future in which war machines of unimaginable capabilities wreak havoc on the remnants of human civilization utilizing hitherto undiscovered forces latent in the ether. Ether physics seemed to many of its practitioners to be firmly settled as the framework within which future progress would take place. The search was on for new understandings of its properties and for further forces that might lie hidden within it.

The breakthrough came at the tail end of 1895, when the relatively unknown Wilhelm Röntgen announced a spectacular new discovery to the world. Röntgen, born in Düsseldorf in 1845, had studied mechanical engineering at Zurich Polytechnic, emerging with a diploma in 1868. Unsure as to his future career, he turned for advice to August Kundt, the professor of physics at the university, who suggested he consider experimental physics as a possibility. Röntgen took the advice and duly produced a doctoral thesis on the study of gases. For the next two decades he rose slowly through the ranks of the German university system. Moving from institution to institution, he was eventually appointed professor of physics at the University of Würzburg in 1888. Like Hertz, Röntgen had been working during the 1880s on the effects of electrical currents in dielectric media like the air. Also like Hertz, he was making a reputation for himself as an adept experimentalist, helped no doubt by his engineering background. In Würzburg during the mid-1890s he interested himself in experimental research on the properties of cathode rays. The result of that interest was his announcement to the world that he had discovered a hitherto completely unknown form of radiation emanating from the Crookes tubes he used for his experiments. Röntgen dubbed the mysterious radiation X rays.

Röntgen's experiments were designed particularly to examine the properties of cathode rays that had leaked from the discharge tube. To this end, in order to shield the fluorescence caused in the tube by the rays, the tube was masked by a covering of black paper. In the course of his experiments, Röntgen noticed a curious phenomenon. A screen coated with barium platinocyanide placed at some distance from the tube—beyond the range of any stray cathode rays—glowed in the dark whenever a current passed through the tube. Puzzled by the strange effect, he carried

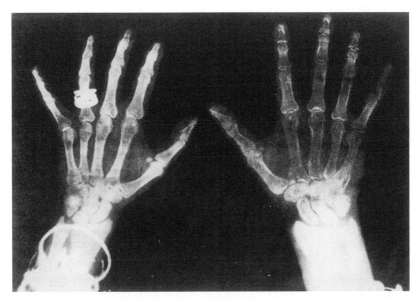

6.5 An X-ray photograph of the hands of the duke and duchess of York in the *Illustrated London News*, July 1896.

out more experiments to try to understand what was going on. Gradually he became convinced that he had discovered a completely new kind of ray—a ray that could pass through solid objects. He even found that he could take photographs with it. A paper announcing his discovery to the world was hurriedly communicated to the Würzburg Physical Medical Society just after Christmas 1895 and copies sent out to eminent physicists throughout Europe. They caused a sensation. According to Röntgen, the rays appeared to behave just like light, except that they could pass through solid objects. His paper included a photograph to illustrate his claims. It was a picture of a human hand—his wife's—made transparent by the rays.

Experimenters rushed to repeat Röntgen's experiment. In England, the electrical engineer A. A. Campbell Swinton was one of the first to succeed, marveling at "the exceedingly curious fact that bone is so much less transparent to these radiations than flesh and muscle, that if a living human hand be interposed between a Crookes tube and a photographic plate, a shadow photograph can be obtained which shows all the outlines and joints of the bones most distinctly"[34] (figure 6.5). In Cambridge,

[34] A. A. Campbell Swinton, *Nature*, 1896, 53: 276.

the physicist J. J. Thomson rushed to investigate the mysterious rays' properties, writing to *Nature* before the end of February with an account of experiments at the Cavendish Laboratory confirming Röntgen's views on the rays' lightlike qualities. In the United States, the flamboyant electrical entrepreneur Thomas Alva Edison quickly recognized the rays' potential, announcing with great fanfare to the press his intention to produce an X-ray photograph of a living human brain. Back in Germany, Röntgen himself performed a demonstration of his discovery for the delectation of the kaiser in person. Experimenters labored hard to improve the technology, finding out what size and shape of tube was best to produce the X rays and trying to understand just what their properties were.

It did not take long for medical men to recognize X rays' potential either (figure 6.6). After all, the first published X-ray photographs were of the bones inside a human hand. The *British Medical Journal* quickly commissioned Sydney Danville Rowland, a young medical student at St. Bartholomew's Hospital in London, "to investigate the application of Röntgen's discovery to Medicine and Surgery and to study practically its applications."[35] Before the end of February he had demonstrated his results before the Medical Society of London. Earlier the same month John Cox, professor of physics at McGill University in Montreal, used X-ray photography to help doctors locate and remove a bullet from a patient's leg. In Cambridge, Edward Douty, surgeon in charge of the gynecology department at Addenbrokes Hospital, took up X rays, while the Cavendish Laboratory provided an informal service to the hospital as well. By May, the first medical journal devoted to X rays had already appeared—the *Archives of Clinical Skiagraphy*. X rays were taken up as therapy as well as for diagnosis. Leopold Freund in Vienna used X rays to remove a mole from a young girl's back. By the early years of the twentieth century, X rays had become a standard part of the armory of hospital electrotherapy departments.

The public was fascinated by X rays and their possibilities. "Roentgen mania," as the *Electrical Engineer* called it, swept across Europe and America. Crowds flocked to X-ray photography studios to have their innards photographed. Edison developed a special fluorescent screen so that people could see their own insides without even having to wait for a photograph to be developed. By 1897 a "Thomas A. Edison X Ray

---

[35]Quoted in M. Weatherall et al., *On a New Kind of Rays* (Cambridge: University Library, 1995), 5.

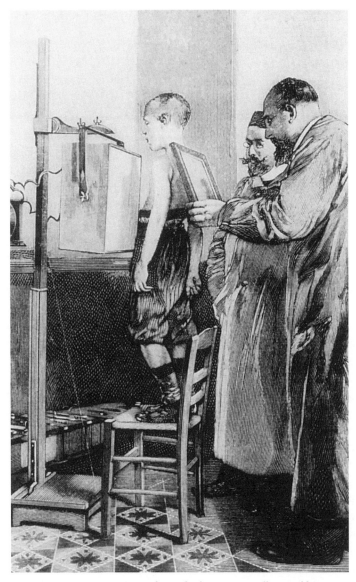

6.6 X rays were soon in use for medical purposes, as illustrated here.

kit" was on the market. Such gimmicks soon became popular fairground attractions. People were particularly intrigued by the possibilities X rays offered of making solid objects transparent. Cartoonists had a field day with the concept, with *Punch* publishing a cartoon depicting the kaiser's discomfiture when John Bull's considerable backbone was revealed by

X ray. It did not take long for the more prurient possibilities to be
explored either:

> I'm full of daze
> Shock and amaze;
> For now-a-days
> I hear they'll gaze
> Thru' cloak and gown—and even stays,
> These naughty, naughty Roentgen Rays.[36]

as one magazine speculated. Newspapers advertised X ray–proof cloth-
ing to protect the public from the prospect of an involuntary scientific
striptease. X rays soon became a staple in futuristic science fiction.

Among physicists, X rays quickly proved to be a particularly fruit-
ful field for new experimental inquiries. Ambitious experimenters could
hope to make their reputations by expanding physicists' understanding of the mysterious rays and their properties, or even by coming up
with the discovery of yet another hitherto unknown kind of radiation.
In France, Henri Becquerel, scion of a distinguished scientific dynasty
stretching back to the beginnings of the century, took up the search with
enthusiasm. He was intrigued by the possibility that the source of X rays
might be the phosphorescence in the glass walls of the Crookes tubes
and set out to investigate whether other sources of phosphorescence also
produced hitherto unknown kinds of radiation. After some false starts,
he found that a sample of uranium salts apparently did just that. In his
report to the Académie des Sciences, Becquerel described how he had
wrapped a photographic plate in sheets of thick black paper to exclude
sunlight and then placed a sample of the phosphorescent substance on
top of the paper. When the photographic plate was developed, the image
of the phosphorescent substance could clearly be seen on the negative. It
seemed that the uranium salts gave off some kind of radiation that could
pass through the black paper and affect the plate.

A few years later, another French physicist, René Blondlot, professor
of physics at Nancy, seemed on the verge of an equally groundbreaking discovery. Like Becquerel, Blondlot had been inspired by Röntgen's
researches to carry out his own investigations. He took up the task of
looking into the physical properties of X rays, being particularly concerned to find out whether or not they could be polarized, like ordinary

---

[36] Quoted in C. Caufield, *Multiple Exposures*, 7.

light. Experimenters had so far been unable to find a way of polarizing X rays—something of a problem for those who argued that they were waves in the ether just like other forms of radiation like light or electric waves. Blondlot suggested that X rays might in fact already be polarized and carried out some experiments, using the brightness of an electric spark as a detector to confirm his suspicions. He soon became convinced, however, that what he was detecting with his spark was not X rays at all but yet another new kind of radiation, which he dubbed N rays, after his native city, as he announced his results to the world in spring 1903. Other experimenters rapidly confirmed Blondlot's discovery and added to it. Particularly intriguing seemed the discovery that N rays were given off by the nervous systems of living beings. The eminent French physicist Arsène d'Arsonval, expert on the interactions of electricity with the human body, described how the rays could be detected emanating from Broca's center in the brain during the process of speech. Some laboratories consistently failed to reproduce the French results, however, notably the illustrious Cavendish in Cambridge. In an effort to understand Blondlot's techniques better, the American physicist R. W. Wood visited his laboratory at Nancy. During one experiment he surreptitiously removed a vital part of Blondlot's apparatus. The N rays continued to appear nonetheless. To his opponents the revelation was decisive. N rays were a figment of Blondlot's imagination. By the end of 1904 they had disappeared back into the ether.

Becquerel's radiation did not go away, however. His discovery became instead the starting point for the researches of a recent Sorbonne graduate who was looking for an interesting topic for her doctoral thesis. Maria Sklodowska, or Marie Curie as she became better known following her marriage to the French physicist Pierre Curie, had come to Paris in search of a physics education in 1891. Determined to stay in France following her graduation and marriage, Marie Curie turned to Becquerel's work as a source for further research at the end of 1897 precisely because little had been done with it since 1896. Everyone else in the field was too busy with X rays to pay much attention to the curious rays given off by uranium. Marie Curie's initial aim was to try and understand the ionizing behavior of Becquerel's radiation—its capacity to make air a conductor of electricity. For this she would use the sensitive measuring apparatus recently devised by her husband. She was soon intrigued, however, as she tested different materials, by the fact that pitchblende—a compound of uranium—seemed to give off more of the mysterious rays than did pure uranium itself. She surmised that this meant the strange radiation was not specific to uranium after all and that pitchblende contained another, hith-erto unknown, element that possessed the mysterious radiating property to an even greater extent.

Pierre soon abandoned his own researches to help his wife, and the two Curies set about identifying the constituent of pitchblende that seemed to give off the new rays in such copious quantities. This was dirty work. Large quantities of pitchblende had to be broken down into its constituent parts to isolate the tiniest amounts of the mystery element that seemed to produce the rays. They eventually concluded that not one, but two different new elements were hidden away in the pitchblende. In July 1898, they announced the existence of a new element they called polonium (after Marie Curie's native Poland) to the Académie des Sci-ences. Their joint paper was titled "On a New Radio-active Substance Contained in Pitchblende." A new word had entered the language of physics. The day after Christmas 1898 witnessed another presentation to the academy. This one was titled "'On a New Strongly Radio-active Substance Contained in Pitchblende." They had found their second element, dubbed radium, and had separate spectroscopic evidence of its existence, provided by Eugène Demarçay (figure 6.7). Marie Curie took upon herself the monumental task of distilling a pure sample of the two elements from the necessary mountains of pitchblende. In a presentation to the International Congress of Physics held in Paris in 1900 to coincide with the universal exposition, the Curies outlined their discovery and their latest work on the properties of the strange rays to a fascinated in-ternational audience. They ended their lecture with a question: was the source of this mysterious energy to be found inside radioactive bodies, or outside them? As they and their audience were beginning to realize, radioactivity seemed to violate some of the most hallowed principles of physics.

In 1903 the Curies, along with Henri Becquerel, were awarded the Nobel Prize for their discoveries. Following her husband's tragic death in a street accident in 1906, Marie Curie devoted her life to radioactiv-ity. From 1907 onwards she presided over Parisian research on the new phenomenon, carefully garnering and protecting access to the difficult-to-acquire sources of the mysterious radioactivity. Her laboratory was devoted in particular to establishing standards in radioactivity, providing accurate measurements of emissions from different sources. As with X rays, much early interest in radioactivity focused on its possible medical uses. If X rays could cure, then radioactivity, which seemed so similar in its effects on the body, could be used in the same way. The telephone

6.7 Marie and Pierre Curie's shed laboratory, where the first samples of the new element radium were isolated.

inventor Alexander Graham Bell suggested that "there is no reason why a tiny fragment of radium sealed into a fine glass tube should not be inserted directly into the very heart of a cancer, thus acting directly upon the diseased material."[37] Outside of physics laboratories, hospitals and medical practitioners made up most of the market for radioactive substances until well into the twentieth century. The stuff was expensive. Laboratory directors such as Marie Curie had to make sure that they forged good contacts with industrial suppliers to make sure that their research was not hampered by their raw material's running out.

X rays and radioactivity provided the ingredients for a major shake-up of physics. Radioactivity in particular seemed to represent an entirely new kind of energy that seemed to defy established categories. It simply was not clear where it fitted into the physics of energy conservation and the ether. Over the next few decades these strange emanations were to provide the tools and building blocks for an entirely novel picture of the physical world and its constituents. Their discovery and the apparent capacities of physicists to manipulate them at will were the focus of immense public interest. They seemed to be only the first in a long line of discoveries just waiting to be made that would transform human destiny. That X rays and radioactivity appeared to hold the key to curing disease was a major attraction as well. They showed just how valuable physics could be when placed at the service of humanity. Their discovery at the dawn of a new century symbolized the prospects for scientific progress that the future held. Radioactivity seemed set to dominate popular perceptions of physics in the twentieth century as electricity had dominated for most of the nineteenth century. It also appeared to provide a rich new mine of discoveries for the increasing numbers of hopeful young physicists just entering an expanding profession.

## Conclusion

Natural philosophy—what became the new discipline of physics—seemed to offer a great deal to thoughtful observers midway through the nineteenth century. It promised an ever deepening understanding of the mysterious forces that governed the Universe. It promised to provide a continually expanding mine of forces and energies that could be put to work, powering an ever growing economy. To its promoters, physics and the capacity it appeared to provide for technological innovation and

[37] Quoted ibid., 26.

unprecedented manipulation of natural powers seemed to be the key to indefinite progress, intellectual, cultural, and economic. It was a given for most of these Victorian commentators that increased knowledge of the workings of the Universe went hand in hand with such progress. Some of them might believe that increasing knowledge was a perfectly respectable end in itself, but few, if any, denied the imperative that such knowledge should be put to work as well. For physics' entrepreneurs, the constant stream of spectacular new discoveries that flowed from new laboratories in Europe and (increasingly) America were an inspiration. Scientific shows and displays of technological marvels underpinned the power of discovery, reminding audiences—if they needed reminding—of just what physics could offer.

The new forces and energies that seemed to appear so effortlessly from the ether during the second half of the nineteenth century were in fact the products of considerable labor and ingenuity. These were the decades during which physics as a discipline was created and consolidated. This was a process that itself required concerted effort. New constituencies had to be persuaded of the benefits that physics could deliver before the laboratories from which these discoveries emerged could be established and supported. Physics' cultural authority—its claims to provide a better way of looking at and understanding the world—did not burst full-grown from Jupiter's head. It had to be argued for. The new discoveries were an important part of this process themselves. They provided hard evidence for the skeptics (of whom there were many) that physics really could deliver the goods. This was one reason why showmanship remained an integral part of physics throughout the century. Physicists had to show their skeptical audience that they had nature under control. Discoveries that could be made spectacularly visible and provide tangible evidence of the action of otherwise unseen forces were central to their success in securing their cultural niche.

7

## Mapping the Heavens

Astronomy rather than natural philosophy was *the* eighteenth-century science. It was the science that set the standard to which others aspired. The phenomenal success of Newton's *Principia* in setting the study of celestial motions on an apparently secure and certain mathematical footing had made astronomy the archetypal science. Newton's triumph in reducing the night sky's complexities to a simple law set the standard for achievement throughout natural philosophy. There were other reasons for astronomy's high status as well. As Europe's maritime nations squabbled over conquered territories in the New World and the Indies, mastery over the seas became crucial. Solving the problem of longitude—to be able to know one's position on the high seas with precision—was essential and astronomy seemed to be the key. Observatories were founded to map the sky with increasing accuracy. Expeditions set out to study celestial phenomena from outposts of empire—with the aim of positioning those outposts more securely on the terrestrial globe. William Herschel's discovery of Uranus demonstrated the power of new telescopes to push back the frontiers of the unseen. Herschel's ambition was to produce a natural history of the heavens—to produce a map of the skies that charted each celestial object and put it securely in its place. Astronomy's high status and evident utility at the end of the eighteenth century meant that it was one of the few sciences to attract substantial state patronage. As such, it was a fruit ripe for the picking to ambitious young radicals as the new century commenced.

In England, astronomers were at the forefront of opposition to Sir Joseph Banks's corrupt and despotic rule (as they saw it) over the Royal Society and English science. Banks had his fingers tightly wrapped around the government's purse strings and astronomers were determined to get a piece of the action. This was what lay behind the foundation of the Astronomical Society in 1820. With George Bidell Airy's appointment as astronomer royal in 1835, the Greenwich Royal Observatory became the focus of an industrialized science. Greenwich was reorganized as a factory system with banks of computers (of the old-fashioned human variety) toiling away producing numbers in industrial quantities. One of the projected functions of Charles Babbage's ill-fated calculating engines was to be the production of astronomical tables. Other observatories throughout Europe emulated Airy's schemes at Greenwich. Far from being ivory towers, nineteenth-century observatories were at the center of industrial culture. With the advent of the electric telegraph they could be linked together as well. Simultaneous telegraphic observations of celestial events across Europe led to ever more accurate determinations of those observatories' place on Earth. As an afterthought, they gave the world the Greenwich time signal as well. These were in many ways the model institutions of nineteenth-century science.

Not only could astronomy provide an accurate map of the Universe, it had the capacity to provide a chart of its history as well. Looking through a telescope at the heavens was like looking back through time. Clues to the origins of the Solar System could be found by searching the night sky. The nebular hypothesis was one of the most popular—and most controversial—constructions of nineteenth-century science. According to the hypothesis, the Solar System in its present state had developed through the operations of natural law from a cloud of gaseous matter just like the nebulae that Sir William Herschel had glimpsed through his telescopes. The nebular hypothesis—the term was coined by the poly-mathic William Whewell—bore the imprimatur of no less a figure than the illustrious Laplace, who had speculated on planetary origins in his *Exposition du Système du Monde*. In the hands of popular expositors such as Robert Chambers and John Pringle Nichol, the nebular hypothesis was a powerful argument in favor of a progressive, evolving Universe. Evolution in the heavens was convincing evidence of evolution on Earth and an inducement to favor social progress as well. Proving (or disproving) the nebular hypothesis was one of the many motives behind Lord Rosse's construction of his massive telescopes in an effort to determine whether nebulae were truly gaseous or not.

While laboratory physicists looked to Europe's observatories as models of how to organize their own fledgling institutions, astronomers picked up on the latest laboratory technologies as well. Astronomers turned in particular to photography and spectroscopy to provide them with new tools to analyze the heavens. Photography, according to its supporters, could provide new standards of objectivity in the representation of celestial phenomena. It replaced the subjective vagaries of the naked eye and the draughtsman's pen with the cold certainties of light and chemistry. It cut out the middleman as well. Astronomers no longer needed to depend on the whims of painters and engravers. With photography, astronomical objects could draw pictures of themselves. Spectroscopy opened up the possibility of resolving heavenly bodies into their constituent elements. The Sun's spectrum could be compared with that produced by any number of terrestrial elements to give a definitive breakdown of its makeup. The introduction of these new technologies meant that astronomy could take place on a tabletop as well as through ever more massive telescopes. It also made what had been a quintessentially observational science subject to the disciplines of experiment. Astronomical laboratories had to be transportable as well. Astronomers trotted the globe with their apparatus, chasing and capturing the latest celestial apparitions wherever they appeared.

One question that ran through nineteenth-century astronomy is still very much alive today. Victorians were fascinated by the possibilities of extraterrestrial life. William Herschel had speculated publicly about the existence of life on the Moon and even in the Sun. He suggested sunspots might be windows in the hot exterior through which might be glimpsed the cool interior where the Sun's inhabitants could be found. For many Victorians this was an issue with real theological significance. Some commentators argued that since there were indisputably other worlds— the Moon, planets, and stars—they must be inhabited since otherwise their creation could have fulfilled no purpose. Others countered that the existence of life on other worlds would rob the Christian revelation of its uniqueness. By the end of the century, the question had moved beyond the abstract. The rise of new communication technologies like wireless telegraphy raised the possibility of actually communicating with the mysterious inhabitants of Mars or the Moon. Science fiction writers such as Jules Verne and H. G. Wells speculated concerning the possibility of travel to the stars. Astronomers such as Percival Lowell and Giovanni

Schiaparelli claimed to have observed canals on Mars—conclusive proof of extraterrestrial life.

Astronomy was transformed during the course of the nineteenth century, setting the standard for collaborative, multidisciplinary science on a massive scale. Observatories had never really been lonely watchtowers where solitary sentinels scanned the skies, and by the nineteenth-century observatories were centers of intensive mass labor. Airy at Greenwich imposed the division of labor as ruthlessly as any Victorian mill owner. Astronomy combined a reputation as the archetypal exact science—mathematical and abstract—with an enviable capacity to draw on state patronage. Astronomy mattered for imperial governance. Astronomy demanded resources on a massive scale as well. Its practitioners needed access to glassmakers, instrument makers, electricians, mathematicians, accountants and bean-counters, and a host of others. Managing an observatory was very much like managing a factory and needed the same kinds of disciplinary surveillance. Men such as Airy and Adolphe Quetelet in Belgium and Johann Franz Encke in Berlin needed to deploy and manage their forces efficiently not just in their observatories but across whole continents to achieve their goals. As astronomers taught physicists valuable lessons about managing large institutions, so physicists provided astronomers with a whole panoply of new resources for studying the stars. The division between the disciplines was increasingly fluid as personnel, practices, and resources passed back and forth between the one and the other.

## Industrial Astronomy

Nineteenth-century astronomy was very far from being the solitary, individualist pursuit of popular imagination. For much of the previous century astronomy had increasingly been recognized as an essential adjunct to maritime supremacy and prowess. State-supported observatories were established in England, France, and the German lands not because of some abstract urge to further knowledge of the heavens but because that knowledge was recognized as having real strategic significance. Accurate maps of the stars were needed to be able to accurately position ships at sea. Astronomy was widely regarded by many—John Harrison and his clocks notwithstanding—as the real key to solving the problem of longitude. In England by the beginning of the century, astronomical endeavors were the subject of substantial state patronage. The British Admiralty, in particular, poured money into projects that were perceived as bolstering the navy's capacities to rule the waves. The primary function of the Royal Observatory at Greenwich, established by Charles II in 1675 with the aim of extending maritime power very much in mind—and other, Continental establishments—was not to make new astronomical discoveries, but to continually hone and refine knowledge of the exact position of known objects in the sky. It was a business that needed routine and careful management as much as innovation.

Partly as the result of the significant amounts of state largesse channeled through the British Admiralty into astronomy, the discipline was one that early attracted the reforming attentions of zealous Young Turks. The Board of Longitude, which under the aegis of the Admiralty oversaw the publication of the *Nautical Almanac* with its charts of astronomical data, was regarded by them as being in hock to the corrupt cronies of Sir Joseph Banks, who, through his long-standing presidency of the Royal Society, maintained a stranglehold over English science. The foundation of the Astronomical Society in 1820 was a direct challenge to Banks's authority—and one that he strenuously opposed. It was an effort to establish an alternative power base on the part of reforming astronomers. The Astronomical Society's reforming stalwarts wanted to have their say in the distribution of the Admiralty's coppers. They saw themselves as disciplined, meritocratic, and vocationally minded gentlemen who could rise to the challenge of putting English astronomy on a proper, businesslike footing. They saw their opponents as amateurish dilettantes, wedded to effete aristocratic interests. Leading members of the Astronomical Society—such as Francis Baily, John Herschel, and Sir James South—played a leading role in efforts to reform the Royal Society following Banks's death in 1820.

Business was the model for the Astronomical Society's cadres of reformers. Many of its founders had close links with the City. Francis and Arthur Baily, along with Benjamin Gompertz, were stockbrokers. Charles Babbage was the son of a wealthy banker. In their view, the foundation of astronomy—like the foundation of good business—was precise measurement and exact calculation: in a word, good bookkeeping. Astronomy, like business, both encouraged and required a habit of exactitude founded on discipline of self and of others. As such, the central function of astronomy was to produce more accurate tables of the skies. Rather than indulging in theoretical speculation, the new society's members were to take on the task of placing their discipline on a sound base of reliable calculation. The key to progress in astronomy as in financial speculation was the elimination of error, and that was best achieved by placing procedures on a proper and transparent footing. As Baily remarked of Babbage's projected calculating engine (and with both accountancy and astronomy in mind), "the great object of all tables is to save time and labour, and to prevent the occurrence of error in various computations."[1] Herschel concurred that good nautical computation "would consist in approximating as nearly as possible to that pursued in the observatory, and divesting it of those technicalities which are not only puzzling to learn, but which really act as obstacles to its improvement by placing it in the light of *a craft and a mystery*."[2]

That ideal was soon in the process of being realized at Greenwich's Royal Observatory, following the appointment of Cambridge senior wrangler George Bidell Airy as astronomer royal in 1835. Like many of the mill owners and Victorian factory managers with whom he had so much in common, Airy was very much a self-made man. Born in 1801, he entered Cambridge in 1819 and graduated as senior wrangler before being elected to a fellowship at Trinity College. His mathematical and astronomical interests aligned him at Cambridge with the Analytical Society's reforming clique—he was a student of the meliorist reforming don (and later bishop of Ely) George Peacock. After a stint in the Lucasian Chair of Mathematics, he was appointed in 1827 as Plumian Professor of Astronomy at the university and director of the recently reestablished university observatory. In a foretaste of things to come, he managed to persuade the university authorities to substantially increase the salary that came with the post. He repeated the trick when he was offered the post of astronomer royal. Unlike his predecessors, Airy did not intend to combine the post with a lucrative ecclesiastical living and demanded the extra cash to make up the difference. He ruled the roost at Greenwich for the next half-century, transforming it into the epitome of nineteenth-century observatories.

Airy imposed a "factory mentality" on the Royal Observatory. Work there was organized according to a strict hierarchy. At the top of the tree, of course, was Airy himself. Beneath him in the chain of command were his trusted lieutenants, Cambridge graduates who looked after the day-to-day management of the institution. Lower in the pecking order were the "obedient drudges"—the computers and observers who did the work. They were typically appointed in their midteens and trained exclusively to carry out particular specialized tasks or calculations. They were

---

[1] Quoted in W. Ashworth, "The Calculating Eye," 415.
[2] Quoted ibid., 431.

selected straight from school on the basis of competitive examination and typically left to become City clerks within ten years. Like contemporary mill owners and other enthusiasts for the factory system, Airy was well aware of the advantages of juvenile labor—it was easily trained, malleable, and above all, cheap. Under Airy's single-minded direction, Greenwich became a veritable production line of astronomical observations and calculations, churned out in published form on a regular and reliable basis. The observatory's brief, in Airy's view, was not "watching the appearances of the spots in the sun or the mountains in the moon, with which the dilettante astronomer is so much charmed . . . it is to the regular observation of the sun, moon, planets, and stars (selected according to a previously arranged system), when they pass the meridian, at whatever time of day or night that may happen, and in no other position; observations which require the most vigilant care in regard to the state of the instruments, and which imply such a mass of calculations afterwards, that the observation itself is in comparison a mere trifle."[3]

Other European observatory managers concurred with Airy's vision of how astronomy should be organized. Indeed, the superior managerial skills of Continental observatory directors was one of the factors English reformers held up as necessitating a thorough overhaul of native practices. The work of the German astronomer Friedrich Wilhelm Bessel was celebrated by Herschel as "the perfection of astronomical bookkeeping."[4] When the Göttingen-educated astronomer Johann Franz Encke (discoverer of the eponymous comet) became director of the Berlin Observatory in 1825, he initiated a thoroughgoing reform of the institution. With the support of the influential Alexander von Humboldt he lobbied successfully for more funds, better instruments, and a new building for his establishment. Under his direction the Berlin Observatory acquired an enviable reputation for the quality and accuracy of its star catalogues. Similarly, François Arago in France and Adolphe Quetelet in Brussels made their reputations as observatory directors largely on the basis of their managerial talents. Like Airy, both instituted regimes at their observatories that sought to replace the idiosyncrasies of individual observers with disciplined, instrumentalized, and routinized procedures. It was no coincidence that Quetelet's other claim to fame was as one of the founders of the science of "social physics." In his statistics, as in his astronomy, the aim was to eliminate variation and cultivate uniformity.

---

[3]G. B. Airy, "Greenwich Observatory," *Penny Cyclopaedia*, 1838, **11**: 442.

[4]Quoted in W. Ashworth, "The Calculating Eye," 429.

The controversy surrounding the disputed discovery of the planet Neptune in 1846 provides an instructive example of the priorities (and nonpriorities) of industrial astronomy. In late 1845, the diffident young Cambridge mathematician John Couch Adams approached Airy with the intriguing suggestion that, by calculating from hitherto unexplained observed perturbations in the orbit of the planet Uranus, he could predict the position of a new and previously unsuspected planet beyond Uranus's orbit. He had already shown his results to James Challis, Airy's successor as Plumian Professor and director of the Cambridge Observatory. Airy ignored Adams's suggestion that a search for the new planet in the predicted position might be a worthwhile proposition. In the meantime, the French astronomer Urbain Jean Joseph Leverrier had been carrying out his own calculations. Unlike Adams, he published his results and communicated his findings to a number of European observatories. On 25 September 1846, two days after receiving his communication, the Berlin astronomer J. G. Galle wrote to Leverrier, "The planet whose position you indicated really exists. The same day I received your letter I found a star of the eighth magnitude that was not recorded on the excellent Carta Hora XXI (drawn by Dr. Bremiker) . . . The observation of the following day confirmed that it was the planet sought."[5]

While Leverrier was lionized across Europe for his discovery, news leaked out in England that Adams had suggested the existence of this planet before Leverrier had, but that Airy and Challis had failed to act on his suggestion. The two men were pilloried in the press as a result. Airy was unrepentant, however. In his view, searching the skies for errant planets was no part of the Royal Observatory's remit. As he pointed out in another context to Greenwich's Board of Visitors, "the Observatory is not the place for new physical investigations. It is well adapted for following out any which, originating with private investigators, have been reduced to laws susceptible of verification by daily observation."[6] The observatory's primary function, he insisted, was the measurement and calculation of astronomical data for purposes of national utility. Intriguing as Adams's calculations might have been, it was not Airy's job to pursue them further. The regime at Greenwich was simply not designed to accommodate such haphazard undertakings. It was for Adams as a private individual to pursue his potential discovery with his own resources; it was not the business of state-sponsored industrial astronomy. Airy had been unimpressed,

[5] Quoted in *Dictionary of Scientific Biography*, s.v. "LeVerrier," 277.

[6] Quoted in A. Chapman, "Private Research and Public Duty," 122–23.

in any case, by the speculative nature of Adams's calculations. They took too much for granted for the hardheaded business astronomer. Too much speculation—in astronomy as in fiscal affairs—was something to be avoided. Like his friend William Whewell, Airy drew a distinction between the progressive sciences—those that were still engaged in the process of discovery—and the permanent sciences that were already fully worked out. In his view, only the permanent sciences (like his brand of astronomy) should be eligible for state support.

More expressive of Airy's views concerning the proper role of the Royal Observatory was his grand plan, developed during the late 1840s and early 1850s, to make Greenwich the central node in an international network of observatories. Using the rapidly developing new technology of electromagnetic telegraphy it would be possible, he argued, to send signals practically simultaneously between different observatories, marking the time at which prearranged observations of particular astronomical phenomena were carried out. The result would be vastly improved accuracy in measurements of those observatories' spatial location and hence in the astronomical data they produced. As the first step in this plan Airy, with the collaboration of Charles Vincent Walker, former secretary of the London Electrical Society, outlined a scheme to hook the Greenwich Observatory into the expanding national telegraph network with the aim of sending out a standardized telegraphic time signal throughout the nation. Airy argued to the Board of Visitors that the telegraph could be "employed to increase the general utility of the Observatory, by the extensive dissemination throughout the Kingdom of accurate time-signals, moved by an original clock at the Royal Observatory."[7] The vision was one where "we may soon expect to see every series of telegraph-wires forming part of a gigantic system of clockwork, by means of which, timepieces, separated from each other by hundreds of miles, may be made to keep exactly equal time, and the clocks of a whole continent move, beat for beat, together."[8]

Airy's and Walker's plan required unprecedented cooperation not only between Greenwich and Continental observatories but among a range of business interests as well. Telegraph and railway companies had to be convinced of the benefits that might accrue from the distribution of telegraphic time from the Royal Observatory. He had to persuade them

[7] Quoted in D. Howse, *Greewich Time and the Longitude*, 95.
[8] G. Wilson, *Electricity and the Electric Telegraph* (London, 1855), 59–60.

that his plan would turn Greenwich time into a universal commodity: "wherever we choose to stretch the telegraph-wires throughout the length and breadth of the land, we could set up a clock and read on its face the evidence of the care which the far distant astronomer bestowed on his observatory clock."[9] For Airy, beyond the virtue of embedding his observatory ever more firmly in the commercial life of the country, the ultimate payoff of the project was the production of an accountable network of observatories throughout Europe. As Walker explained to the *Times*, with the signaling system in place, "Mr. Airy at Greenwich, and M. Arago at Paris, will thus be able to fix a time when the eye of each shall be directed to the same star at the same time, and signal to each other as each wire [of the transit instrument] is passed."[10] In the United States, observatory managers such as William Cranch Bond at Harvard College Observatory and Truman Stafford at Chicago's Dearborn Observatory offered their commercial services in selling the true time to local jewelers and railway companies.

The establishment of the Greenwich time signal provides a fine example of the ambitions of industrial astronomy in action. Early nineteenth-century reformers, in England at least, regarded the state of astronomy as parlous. The science had been hijacked by a gaggle of effete, ineffective, and self-interested dilettantes who lacked the discipline to set astronomy on a proper footing and failed to appreciate its possibilities. Thus, astronomy's institutions required a thorough overhaul by hardheaded business astronomers who could impose the regulation the science needed. Observatories were to be regarded as factories dedicated to the production of numbers in industrial quantities. For fans of Adam Smith, it was self-evident that the best way of maximizing production in a factory was by a ruthless imposition of the division of labor. For astronomers such as Airy in England and Quetelet in Belgium, exactly the same lessons pertained to the conduct of observational astronomy. It was best carried out by hierarchically ordered and disciplined cadres of workers organized according to the division of labor. By midcentury, therefore, astronomy was a model of coordinated and collaborative science. In the interests of uniformity, observatory managers turned more and more to instrumentation and strict regimes of calculation and observation in order to minimize the impact of individual idiosyncrasies on their intellectual productions.

[9] Ibid., 63–64.
[10] Quoted in I. R. Morus, "The Nervous System of Britain," 466.

## The Nebular Hypothesis

Other forms of astronomy that also fitted in well with Victorian ideals of progress developed during the first half of the nineteenth century. Following William Herschel's lead, astronomers scanned the sky with ever larger and more powerful telescopes. New astronomical discoveries—finding novelties in the night sky—were sources of considerable kudos. William Herschel, after all, had made his name and practically founded an astronomical dynasty with his spectacular discovery of Uranus. Telescopic astronomy gained additional significance during the first half of the nineteenth century as general perceptions of the Universe transformed. Eighteenth-century cosmologists typically regarded the cosmos as a timeless, changeless equilibrium. Their nineteenth-century inheritors more frequently visualized the Universe as being in a state of continuous progression. Telescopes that made it possible to gaze ever further into the cosmic vastness could therefore be regarded as doing far more than just providing a glimpse of the Universe's structure. They opened a window onto the Universe's past as well. Discussions on such topics had massive contemporary resonance. If the Universe was in a state of evolution, then maybe so was life on Earth. If progress in nature was a matter of natural law, then maybe social progress and emancipation rather than subservience to the status quo should be the norm as well.

One theory held particular resonance. The nebular hypothesis, as it was popularly dubbed, had an impressive pedigree. It was partially founded on William Herschel's compendious observations of nebulae—what seemed to be clouds of gaseous matter in space (figure 7.1). It carried the hallmark of Newtonian authenticity provided by the authority of the French physicist Pierre-Simon Laplace. As Herbert Spencer, one of the nebular hypothesis's many enthusiastic promoters, argued, "To have come of respectable ancestry is *prima facie* evidence of worth in a belief as in a person; while to be descended from discreditable stock is, in one case as in the other, an unfavourable index."[11] By that criterion, the nebular hypothesis came from distinguished stock indeed. Laplace had suggested in his *Exposition du Système du Monde*, at the close of the previous century, that nebulae might be regarded as the birthplaces of the stars and planets. In strict accordance with Newtonian mechanics, he envisaged a process whereby swirling clouds of cosmic gas gradually coalesced, first into clumps of matter around a slowly thickening central

---

[11] Quoted in S. Schaffer, "The Nebular Hypothesis and the Science of Progress," 132.

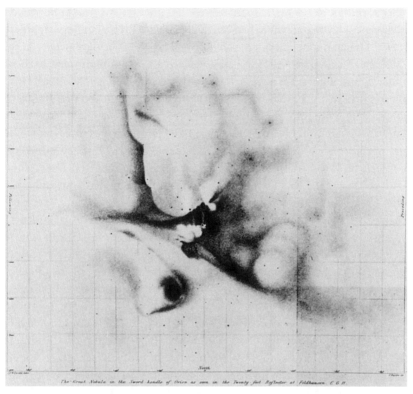

The First Nebula in the Sword-handle of Orion as seen in the Twenty-foot Reflector at Feldhausen. C G H

7.1 The Orion nebula as pictured by John Herschel.

mass, and finally into discrete satellites orbiting that glowing mass—just like the planets orbiting around the sun. To many, it seemed a persuasive scenario. Auguste Comte, the rising star of French philosophers, popularized and defended the idea as part of his developing creed of positive philosophy.

Nebulae came in all shapes and sizes. John Herschel in his popular volume *Treatise on Astronomy* for the best-selling Cabinet Cyclopaedia unreservedly credited his father for "the most complete analysis of the great variety of those objects which are generally classed under the head of Nebulae, but which have been separated by him into—1st, Clusters of stars, in which the stars are clearly distinguishable; and these, again, into globular and irregular clusters. 2d, Resolvable nebulae, or such as excite a suspicion that they consist of stars, and which any increase of the optical power of the telescope may be expected to resolve into distinct stars; 3d, Nebulae, properly so called, in which there is no appearance whatever of stars; which, again have been subdivided into subordinate classes, according to their brightness and size; 4th, Planetary nebulae; 5th, Stellar nebulae; and, 6th, Nebulous stars."[12] The "great power" of William Herschel's telescopes had revealed "an immense number of these objects, and shown them to be distributed over the heavens, not by any means uniformly, but, generally, speaking, with a marked preference to a broad zone crossing the milky way nearly at right angles."[13] About his father's speculations John Herschel was rather more circumspect. "The nebulae furnish," he remarked, "an inexhaustible field of speculation and conjecture." Most of them were made up of stars, he asserted, but "if it be true, as, to say the least, it seems extremely probable, that a phosphorescent or self-luminous matter also exists, disseminated through extensive regions of space . . . what we may naturally ask, is the nature and destination of this nebulous matter? Is it absorbed by the stars in whose neighbourhood it is found, to furnish, by its condensation, their supply of light and heat? or is it progressively concentrating itself by the effect of its own gravity into masses, and so laying the foundation of new sidereal systems or of insulated stars?"[14]

Some speculations concerning the Universe's origins could have distinctly subversive implications. The radical lecturer Thomas Simmons Mackintosh—a committed disciple of the utopian socialist Robert Owen—made quite a name for himself during the 1830s across Britain with his electrical theory of the universe. Lecturing in Owenite Halls of Science and Mechanics' Institutes the length and breadth of the country, Mackintosh put forward a view of the Universe that had electricity rather than gravity as its driving force—all "motion throughout the solar system is effected by the agency of electricity."[15] He took advantage of the reappearance of Halley's comet in 1835 to promote his theory, arguing that comets were "immense volumes of aeriform matter discharged from the sun by the agency of electricity"[16] and that they would eventually condense into planets. Electricity acted to prevent the planets from falling into the sun, but as that electricity dissipated, the eventual fate of the Solar System was inescapable: "The river flows because it is running

[12] J. Herschel, *Treatise on Astronomy* (London, 1833), 40.

[13] Ibid., 401.

[14] Ibid., 406–7.

[15] T. S. Mackintosh, "Electrical Theory of the Universe," *Mechanic's Magazine*, 1835–36, **24**: 228.

[16] Ibid., 11.

down; the clock moves because it is running down; the planetary system moves because it is running down; every system, every motion, every process, is progressing towards a point where it will terminate."[17] The relentless unfolding of natural law meant that the universe had a discrete beginning and end. There was no room for God in Mackintosh's cosmological picture and no room for Christian salvation either. The radical message behind his cosmology was that mankind needed to make its own salvation on earth while there was still time.

Even more dangerous in the eyes of many gentlemen of science, however, was the use made of the nebular hypothesis in the notorious *Vestiges of the Natural History of Creation.* Published anonymously in 1844, *Vestiges* was in fact the work of the Edinburgh publisher Robert Chambers. Born in 1802, the son of a hand-loom weaver, Chambers had by the 1830s made a name and some fortune for himself as a bookseller and publisher. Along with his brother William he ran *Chambers's Edinburgh Journal* from the early 1830s onwards, selling eighty thousand copies a week by the 1840s. Originally a Tory, Chambers was by now a firmly liberal Whig and the journal an expression of middle-class Whig values of improvement and progress. *Vestiges* was dangerous in the eyes of its opponents precisely because it was aimed at and written for exactly the kind of solid, respectable middle-class citizen who read *Chambers's Edinburgh Journal.* It could not be dismissed as easily as the patently radical rantings of an avowed socialist such as Mackintosh. The anonymous author (though many among the gentlemen of science suspected Chambers by the end of the 1840s) had done his homework as well. *Vestiges* made good use of the latest in natural philosophy to underpin its message.

Chambers's argument was simple. The history of the universe was the history of the gradual unfolding of natural law. He hammered the lesson home with examples ranging from the supposed development of stars and planets from nebular gas to the transmutation of species. For cautiously reformist gentlemen of science this was dangerous stuff. "The nebular hypothesis," *Vestiges* announced to its readers, "is, indeed, supported by so many ascertained features of the celestial scenery, and by so many calculations of exact science, that it is impossible for a candid mind to refrain from giving it a cordial reception, if not to repose full reliance upon it, even without seeking for it support of any other kind ... seeing in our astral system many thousands of worlds in all stages of formation, from the most rudimental to that immediately preceding the present condition of those

---

[17] T. S. Mackintosh, *The Electrical Theory of the Universe* (Boston, 1846), 371.

we deem perfect; it is unavoidable to conclude that all the perfect have gone through the various stages which we see in the rudimental."[18] The conclusion was inescapable: "the whole of our firmament was at one time a diffused mass of nebulous matter, extending through the space which it still occupies. So also, of course, must have been the other astral systems. Indeed, we must presume the whole to have been originally in one connected mass, the astral systems being only the first division into parts, and solar systems the second." There was another conclusion as well: "that the formation of bodies in space is *still and at present in progress*."[19]

The importance of this appearance of continuous progress (and its implications for the progress of society) had also already been emphasized by John Pringle Nichol—another fan of the nebular hypothesis. "In the vast Heavens, as well as among phenomena around us, all things are in a state of change and PROGRESS,"[20] he proclaimed, making quite explicit the intimate connection between celestial physics and social dynamics. An avowed political radical, Nichol progressed himself during the 1830s from Scottish schoolmaster through popular lecturer and journalist on political economy and natural philosophy until in 1836 he was appointed professor of astronomy at Glasgow. In his popular and influential *Views of the Architecture of the Heavens* (1837) he was adamant that the nebular hypothesis demonstrated the existence of a progressive order in the universe that stretched from the formation of stars and planets to the actions of men on Earth. According to Nichol and his allies, the nebular hypothesis demonstrated the need for and indeed the inevitability of social reform. Society was subject to exactly the same kind of progressive forces that made the Solar System out of an inchoate cloud of stellar gas. Nichol's friend and political ally John Stuart Mill concurred and thought that Nichol's book would be the making of him.

For Mill, Nichol, and their associates, the nebular hypothesis was only one part—though an absolutely crucial part—of a wide-ranging and comprehensive science of progress. The science of progress incorporated Mill's logic, David Ricardo's political economy, George Combe's phrenology, and Jean-Baptiste Lamarck's theory of evolution by means of transmutation as well. Nichol was relentless in his campaign to underpin progressive change in society with the revelation of continuous progress in the heavens. Like Chambers, he emphasized that the nebular

[18] [R. Chambers], *Vestiges of the Natural History of Creation* (London, 1844), 19–20.

[19] Ibid., 20–21.

[20] J. P. Nichol, *Views of the Architecture of the Heavens* (Edinburgh, 1837), 206.

7.2 Lord Rosse's telescope.

hypothesis's main selling point was its very simplicity. Regardless of any cavils from the gentlemen of science, the hypothesis was so compelling as to be self-evident. Anyone with eyes to see and sufficiently power-ful telescopes to look through could confirm the fact for themselves. The nebulae were there to be seen in the heavens and common sense did the rest. For those who could not gain access to large telescopes for themselves, Nichol's crowd-pleasing lectures were an equally compelling substitute. His shows were famous—to some notorious—for their "gor-geous style, gigantic diagrams and enthusiasm." Nichols was celebrated as "the prose laureate of the stars"[21] and traveled as far afield as New York to give his performances. To his opponents, however, his lectures seemed full of bombast rather than substance; the nebular hypothesis a misreading of the evidence of the heavens.

Working out just what the evidence of the heavens might be in this regard was one of the issues the aristocratic Lord Rosse hoped to resolve with the construction of his gargantuan telescope, the Leviathan of Parsonstown, at his family seat at Parsonstown in King's County (now County Offaly), Ireland (figure 7.2). Rosse was an enthusiastic

[21] Quoted in S. Schaffer, "The Nebular Hypothesis," 150.

astronomer with the leisure and resources to indulge his passion on a
massive scale. Building on William Herschel's telescopic achievements,
by the early 1840s Rosse had already constructed a thirty-six-inch reflecting telescope at Parsonstown and was about to embark on an even
more audacious project. Between 1842 and 1845 work was under way
on the Leviathan, which was to be seventy feet in length with a mirror
six feet in diameter. The engineering achievement alone was widely celebrated as a symbol of progress. Rosse "had no skilled workmen to assist
him. His implements, both animate and inanimate, had to be formed by
himself. Peasants taken from the plough were educated by him into efficient mechanics and engineers."[22] Rosse mechanized the process, using
steam-powered tools to polish the gigantic mirror. The final product was
definitional of the aim of telescopic astronomy, and with it Rosse set out
to scour the skies for nebulae and try whether his telescope's awesome
power could resolve them. The more nebulae that could be resolved into
constituent stars, the less plausible was the prospect of what Herschel
had called "true nebulae" and therefore the nebular hypothesis.

Rosse's main ally was Thomas Romney Robinson, an Anglican divine
and director of the Armagh Observatory in Ireland. He was a fervent
opponent of Papism on the one hand and the materialist radicalism associated with the nebular hypothesis on the other. Their target was the
Orion nebula that Nichol had pinpointed in *Views of the Architecture of
the Heavens* as a likely candidate for true nebula status. Early in 1846,
Rosse wrote to inform Nichol that the Leviathan had been successful in
resolving the Orion nebula. It was, as Nichol ruefully remarked, no more
than a "SAND HEAP of stars."[23] Robinson and Rosse had been working
hard to discredit William Herschel's observations, on which the claims
of the nebula's unresolvability—and hence the plausibility of the nebular hypothesis—rested (figure 7.1). Robinson in particular went out
of his way to undermine Herschel's reputation, asserting that his nebular observations were worthless since Herschel was an incompetent
telescope maker. Despite the Leviathan's reputation, however, its observations could not be decisive. Faced with the apparent resolvability of
the Orion nebula, Nichol enthusiastically picked up on others of Rosse's
observations—such as his reports on spiral nebulae—as new evidence
of the existence of the kind of interstellar gaseous fluid required by the
nebular hypothesis.

---

[22] A. Clerke, *A Popular History of Astronomy in the Nineteenth Century* (London, 1900), 115.

[23] Quoted in S. Schaffer, "The Leviathan of Parsonstown," 214.

Disputes concerning true nebularity dragged on for most of the rest of the century. No observations, not even ones with as powerful an instrument as Rosse's Leviathan, could ever be really decisive since so much rested on their interpretation. As the astronomer Otto Struve said to Rosse in 1868, "In my opinion if a nebula is resolvable it will offer the same appearance on any occasion when the images are sufficiently favourable. Thus admitted, your own observations show that, with regard to the central part of the nebula of Orion, this term ought not to be applied, for in different nights you see resolvability in different parts of the nebula."[24] Where Rosse saw stars, Struve and others saw constantly changing patterns of "nebulous matter" sometimes tangling into "separate knots." Rosse and his workers insisted that the superiority of their instrument should give them the final say. There was nothing about their observations that could force their opponents to capitulate, however. Too much was at stake as far as Nichol and his fellow proponents of the science of progress were concerned. Once in the public gaze, Rosse's observations ceased to be his property and could be read in a variety of ways quite conducive to the nebular hypothesis. New kinds of instruments and new technologies were competing with the Leviathan of Parsonstown as well for the status of being the ultimate arbiters on the matter.

## The Sky's Laboratories

While Lord Rosse and his Leviathan might at first sight conform comfortably enough with the traditional image of the astronomer as lonely and heroic watcher of the skies, his operations in reality were quite different. The Leviathan was very much a product of industrial culture. Other innovations in astronomy during the middle parts of the nineteenth century also owed much to the physicist's laboratory and the mechanic's workshop. These decades saw the introduction of a number of new technologies into the practice of astronomy with the aim, at least in part, of making its processes less subject to the vagaries of the human observer and therefore more "objective." Establishing that objectivity was by no means a straightforward task. It was not at all obvious to contemporary commentators that the replacement of human illustrators with photographs as means of recording the appearance of the heavens, for example, was necessarily to make the representations more objective. They had to be persuaded. Turning the observatory into a physicist's

[24]Quoted ibid., 221.

laboratory was a business that involved the introduction of novel kinds of discipline and work organization as well as of new kinds of instruments. Astronomy's audiences needed to be persuaded as well that these novelties really would make their picture of the heavens more real.

Photographic pioneers were quick to suggest that astronomy was one science that might clearly benefit from their services—unsurprisingly given that some astronomers, notably John Herschel, played key roles in developing new photographic technologies. Herschel had been involved in experiments on the sensitivity of various chemical substances to light since the 1820s. He was a close associate of William Henry Fox Talbot, who had introduced the calotype method of photography. Herschel produced his first photographs in 1839 and was instrumental in introducing the term "photography" to describe the new technology. François Arago, the director of the Paris Observatory, was similarly instrumental in bringing the discoveries of the French inventor Louis Jacques Mandé Daguerre to the attention of the Académie des Sciences. He was also one of the first to suggest, shortly after Daguerre's announcement of his new photographic—or daguerreotype—process in 1839 that the new technique might have a useful role to play in astronomy. An early daguerreotype of the moon by Daguerre was probably taken at Arago's suggestion. Photography was touted as a means of making astronomical observations more objective—freeing them from the constraints of human subjectivity. They might also have the virtue of superior sensitivity. Chemicals that reacted to light that the human eye failed to register could capture images of stars and celestial phenomena that mere men might miss. This was one way in which photographs could help with the nebular hypothesis—they might provide evidence of that elusive nebular fluid that the naked eye might miss.

The possibilities of astronomical photography received a major boost when the Harvard astronomer William Cranch Bond exhibited daguerreotypes of the moon taken at the Great Exhibition in 1851. They were not the first of their kind, but they were notable for their clarity and their obvious affinity to naked eye impressions. They were celebrated as showing that photography really could replace and improve upon individual perceptions. Bond had produced the daguerreotypes with the help of Boston photographer J. A. Whipple. It was a process that required considerable experimentation to find the best chemicals for the exposure and a great deal of human ingenuity. Well into the 1850s George Phillips Bond (William Cranch Bond's son and successor as director of the Harvard Observatory) was still emphasizing the extent to which astronomers remained dependent on the skills of artists, engravers, and photographers to produce credible images of celestial objects. Such remarks underlined the difficulties to be overcome in using photography to turn astronomy into what George Bidell Airy called a "self-acting" science. Behind the scenes, photography remained very far from self-acting—it was dependent on its aficionados.

In Britain, one of those aficionados was Warren de la Rue, who for much of the nineteenth century was firmly established as the country's premier astronomical photographer. Born in London in 1815 and educated in Paris, de la Rue was the son of the founder of a stationery manufacturing firm. A member of the London Electrical Society, he was a keen enthusiast for natural philosophy, publishing on chemistry and electricity from the late 1830s. During the 1840s he turned his attention to astronomy and was soon an advocate of photography. In 1851 he produced his first photograph of the Moon, using the newly developed wet collodion process and pointing his camera through his own thirteen-inch reflecting telescope. Coming from a well-heeled manufacturing family had distinct advantages. De la Rue could afford his own private observatory. He had access to the resources (and the leisure) that remained essential for working with a still cumbersome and time-consuming new technology. Exposure times—even for a bright object like the Moon (one of the reasons de la Rue chose it as the object of his early experiments)— were a significant factor. De la Rue had to find ways of accurately tracking his target during the process.

De la Rue worked hard to improve the process, designing a driving clock that could move his telescope in tandem with the Moon. With the new technology—and a move to cleaner air outside London at what was then the picturesque village of Cranford (now a suburb of London)—he could produce images of impressive clarity. He also managed a stereoscopic (three-dimensional) portrait of the Moon that turned the eminent John Herschel into an instant fan. "I hasten to testify my admiration of this transcendent and wonderful effort," the astronomer enthused. "It is a step *in nature* but beyond *human* nature as if a giant with eyes some thousands of miles away looked at the moon through a binocular."[25] From the mid-1850s onwards he was also attracting Airy's attention, always on the lookout for ways of making the observer redundant. Airy argued that photography could introduce new standards of accuracy and objectivity from physics into astronomy: "it will supersede hand-drawing

[25] Quoted in H. Rothermel, "Images of the Sun," 144.

altogether, and even now the results obtained are much more accurate than anything hitherto done by mapping or hand-drawing."[26] Photography could eradicate the otherwise ineradicable "personal equation"—the idiosyncrasy that marked even the most alert and dedicated watcher's observations with an indelible individual taint. As such this import from physics could be an ideal addition to Airy's industrialized Greenwich regime.

Having been involved in studying sunspots during his self-imposed exile at the Cape of Good Hope, John Herschel, inspired by de la Rue's successes with lunar photography, was anxious for experiments with solar photography as well. In 1854 he prompted Edward Sabine—a fellow scientific reformer, campaigner for the "magnetic crusade" to map terrestrial magnetism, and influential member of the British Association for the Advancement of Science's Kew Observatory Committee—that it would be "an object of very considerable importance to secure at some observatory . . . daily photographic representations of the sun, with a view to keep up a consecutive and perfectly faithful record of the history of the spots."[27] De la Rue was soon recruited for the job and duly set up shop at Kew—where the British Association for the Advancement of Science maintained its observatory—with the aid of a £150 grant from the Royal Society. By 1858, de la Rue had perfected a working photoheliograph and instituted a successful regime "to determine all the data necessary for ascertaining the relative magnitudes and positions of the sun's spots."[28] The instrument's possibilities as a means of providing reliable representations of another solar phenomenon—the eclipse—were soon spotted. In 1860 plans were mooted to carry the photoheliograph to Spain to capture an eclipse of the sun there.

Pictures of solar eclipses were notoriously unreliable and hopes were high that photography might prove to be the answer. Illustrators and engravers during the 1830s and 1840s had developed an array of techniques in efforts to make sufficiently realistic representations of the elusive phenomenon. The comparative rarity of eclipses was one problem. Another was their short duration and the difficulty of looking at them for extended periods. Astronomers gave their draftsmen detailed instructions as to what they should look for and how they should try to depict their impressions. Finished drawings were out of the question. Observers made rapid sketches while an eclipse was in progress and tried to fill in the blanks from memory. The camera made such skills

---

[26]Quoted ibid., 145.    [27] Quoted ibid., 152.    [28] Quoted ibid., 153.

redundant, but introduced a whole range of new techniques instead. Photographing an eclipse was a labor-intensive process. Plates had to be carefully prepared beforehand. One assistant stood ready to hand plates to the photographer while another stood by to uncover and cover the telescope at the crucial moments. Another was ready to rush the exposed plates away to be developed immediately—the plates would spoil quickly if not dealt with on the spot. All this took place as a rule in some foreign clime far removed from the astronomer's home observatory.

Expeditions were essential to capture rare and transient astronomical phenomena like eclipses. Astronomers had to pack their bags and move lock, stock, and barrel to the appropriate spot on the Earth's surface where the phenomenon might best be observed. De la Rue's expedition to Spain in 1860 with the Kew photoheliograph is a good example. To deal with the expected difficulties of photographing the elusive phenomenon and the idiosyncrasies of the apparatus, de la Rue had a complete photographic observatory built for the occasion, divided into one part with a removable roof containing his heliograph and another equipped as a photographic room. The Admiralty, which was bankrolling the expedition, put up a ship to transport the astronomers and their traveling observatory en masse to Spain. The resulting photographs were a hit. Detailed preparations were similarly essential for the projected expedition to observe the Transit of Venus across the Sun's face in 1874—an event that had last taken place more than a century previously (figure 7.3). Airy and de la Rue were discussing plans for photographs as early as 1868. A model set up at Greenwich was used to train observers before they set out for the five observing stations in Egypt, Hawaii, New Zealand, and two South Pacific islands. The observing stations and their equipment were all identical, having been built at Greenwich before being shipped out to be reassembled on site. Despite all efforts, however, the expedition was a flop. The results turned out to be wildly inconsistent and nothing from the photographic parts of the enterprise ever saw print. It was an instructive lesson in the limits of instrumentalized standardization.

Spectroscopy was the other addition to the armory of astronomy during the second half of the nineteenth century. The new technique was the outcome of early nineteenth-century observations that the color of flames or of electric sparks between electrodes varied according to the makeup of the electrodes. The light when viewed through a prism gave a spectrum unique to each particular element. The brilliant German optical instrument maker Josef von Fraunhofer had noted that light from the Sun exhibited characteristic lines in its spectrum when viewed through a

7.3 Astronomers testing their equipment in preparation for observing the Transit of Venus.

prism. He had used these lines, which came to be known as Fraunhofer lines, to demonstrate the quality of his optical apparatus. Such a technology could, however, also be used by astronomers to identify the makeup of celestial bodies by studying the characteristics of their light. As James Clerk Maxwell noted, it was a striking vindication of the universality of physics: "when a molecule of hydrogen vibrates in the dog-star, the medium receives the impulses of these vibrations; and carrying them in its immense bosom for three years, delivers them in due course, regular order and full tale into the spectroscope of Mr Huggins at Tulse Hill."[29] The result, according to Mr Huggins himself—William Huggins, owner of a private observatory—was that "an astronomical observatory began, for the first time, to take on the appearance of a laboratory. Primary batteries, giving forth noxious gases, were arranged outside one of the windows; a large induction coil stood mounted on a stand on wheels so as to follow the positions of the eye end of the telescope, together with a battery of several Leyden jars; shelves with bunsen burners, vacuum tubes and bottles of chemicals, especially of specimens of pure metals, lined its walls."[30]

The German physicist Gustav Robert Kirchhoff took the lead in solar spectroscopy with experiments in the 1850s and 1860s. Kirchhoff

[29]J. C. Maxwell, *Scientific Papers of James Clerk Maxwell* (Cambridge, 1890), 2: 322.
[30]Quoted in S. Schaffer, "Where Experiments End," 268–69.

claimed that on the basis of detailed spectroscopic analysis of the Sun's light he and his coworkers could actually reproduce the solar atmosphere in their laboratory. Kirchhoff in 1861 used his spectral findings to promulgate a new theory of the Sun's composition. The Sun, according to his spectroscope, was made up of an incandescent, luminous fluid. Sunspots were cloudlike spots that floated high above its surface. Kirchhoff's model, and his assertion that the Sun's constitution could be reproduced under laboratory conditions, was strongly disputed by the French astronomer Hervé Faye, who argued that drawing such analogies between laboratory experiments and inaccessible celestial phenomena was tendentious at best. Kirchhoff's sunspot model was just as strongly disputed across the Channel in England. The dominant model there—formulated by William Herschel and strongly defended by his son—was that sunspots were holes through the Sun's atmosphere to the dark (and habitable) surface beneath. Unlike the Frenchman, however, British astronomers such as William Huggins and Norman Lockyer were more than happy to agree with Kirchhoff's claim that spectroscopy was the key to understanding the solar (and stellar) constitution.

There was little doubt in Lockyer's mind that spectroscopy was the key to unlocking the Sun's secrets. "There is an experiment by which it is perfectly easy for us to reproduce this artificially," he said of his claim that observed changes in the width of lines in the solar spectrum were the result of pressure changes in the Sun's atmosphere: "we can begin at the very outside of the Sun by means of hydrogen, and see the widening of the hydrogen lines as the Sun is approached; and then we can take the very Sun itself to pieces."[31] It was a powerful claim. One reason for spectroscopy's power as a tool for astronomers was the way it allowed them to make their observations public. Stellar phenomena invisible to the layperson could be reproduced in the laboratory—and more importantly in the lecture theater—and made accessible to all. Astronomers could tell stories that explicitly linked what was going on in the physics of a piece of terrestrial demonstration apparatus to what took place in the heart of the Sun, or the uncharted depths of interstellar space. Spectroscopy and solar physics could be marketed as tools that provided vital information about the age of the Sun and the lifespan of the Cosmos—topics that mattered to a generation obsessed with degeneration, evolution, and the heat death of the Universe.

[31] Quoted ibid., 283.

Victorian astronomers were well aware of the extent to which the importation of these new technologies transformed their discipline. David Gill, Britain's resident astronomer at the Cape Colony, represented them to a Royal Institution lecture audience as the acme of astronomy: "It is these after all that most appeal to you, it is for these that the astronomer labours, it is the prospect of them that lightens the long watches of the night."[32] While their products might be inspirational, the procedures and disciplines accompanying these new practices—carried over from the nineteenth century's growing laboratory culture—had their closest affinities to the industrialized astronomy of George Bidell Airy and his cohorts. As Airy's style of stargazing required the mobilization of ranks of ordered, disciplined observers and computers, astronomical physics needed the mass orchestration of new resources, skills, and workers as well. This was an astronomy that needed chemists, electricians, and photographic entrepreneurs as well as opticians and telescope makers. Ways had to be found of integrating these newcomers into older ways of doing things. Despite his enthusiasm, it was notable that even Airy balked at bringing too much of the new physics into the Royal Observatory, though he did establish an Astro-Photographic and Spectroscopic Department in 1874—it was a tool of spectacular discovery ill-suited to Greenwich's more utilitarian remit. In Britain at any rate, astronomical physics found its feet in smaller, often private observatories like those of de la Rue at Cranford and William Huggins at Tulse Hill, rather than under the aegis of state astronomy.

## Other Worlds

One of the reasons the Victorian public, as well as astronomers, were so fascinated by the prospect of finding out more about the physical characteristics of celestial bodies was that it seemed to shed light on the possibility that those other worlds might be inhabited. The question of extraterrestrial life was both controversial and topical throughout the nineteenth century. It was not just an issue confined to the margins of cultural and intellectual life—some of Europe and America's most respected astronomers lined up to opine on the matter. Neither was this a new issue. By the beginning of the nineteenth century the question of extraterrestrial life and the possibility of a "plurality of worlds" had

[32] Quoted ibid., 267.

a long history. The infamous Giordano Bruno had been an enthusiastic proponent of the idea that life existed not only on the Moon and other planets, but on the stars as well. Johannes Kepler had been a little more circumspect but still believed that other planets were probably inhabited. Even the great Sir Isaac Newton expressed the view that if all parts of the Earth were inhabited, then there seemed no reason to suppose that God would have left the heavens uninhabited. Newton's remarks underlines the theological implications of discussions of the plurality of worlds—implications that were still there in the nineteenth century.

William Herschel—whose reputation as an astronomer remained high throughout the nineteenth century—was a particularly enthusiastic advocate of the plurality of inhabited worlds. He was adamant that the Moon must be occupied by inhabitants of one kind or another. He was confident as well that his increasingly powerful telescopes would eventually provide irrefutable proof on the matter. He even toyed with the idea that he had already seen such evidence. On one occasion he noted following some telescopic observations of the moon that "I believed to perceive something which I immediately took to be *growing substances*. I will not call them Trees as from their size they can hardly come under that denomination . . . My attention was chiefly directed to Mare humorum, and this I now believe to be a forest."[33] Some of his contemporaries thought that Herschel's preoccupation with lunar (and solar) life rendered him "fit for bedlam." He was not, however, the only Enlightenment astronomer to entertain the possibility of extraterrestrial life by any means. Laplace discussed the possibility in his *Mécanique Céleste*. The argument was supported by Jérôme Lalande, professor of astronomy at the Collège Royale, who argued that "the resemblance is so perfect between the earth and the other planets that if one admits that the earth was made to be inhabited, one cannot refuse to admit that the planets were made for the same purpose."[34]

The German astronomer Franz von Paula Gruithuisen was one of the early nineteenth century's most enthusiastic advocates of extraterrestrial life—as well as being one of the century's most prolific astronomical writers. In 1824, in his "Entdeckung vieler deutlichen Spuren der Mondebewohner, besonders eines collassalen Kunstgebäudes derselben" (Discovery of Many Distinct Traces of Lunar Inhabitants, Especially One

[33] Quoted in M. Crowe, *The Extraterrestrial Life Debate*, 63.
[34] Quoted ibid., 79.

of Their Colossal Buildings) he argued that the colored tints he observed on the Moon's surface should be interpreted as evidence of vegetation. He claimed to have seen pathways through his telescope, demonstrating the existence of lunar animals roaming on the surface. He had also seen a variety of geometrically shaped features that he speculated might be artificially constructed roads and cities. One large, star-shaped structure in particular was labeled a temple. Some of his contemporaries regarded all of this as evidence that the professor of astronomy at the University of Munich had—like William Herschel—taken leave of his senses. The plurality of worlds was nevertheless part of the common discourse of astronomical debate in the German lands. Even those such as Carl Friedrich Gauss, Wilhelm Olbers, and Johann Joseph von Littrow, who regarded Gruithuisen's claims as patently absurd, were themselves sympathetic to the possibility that life existed on other worlds.

In Britain, William Herschel's stellar reputation, if nothing else, guaranteed discussions of the plurality of worlds a sympathetic hearing. John Herschel, as an assiduous defender of his father's achievements, was an advocate of pluralism as well, if a rather more circumspect one than his parent. The planets, according to John Herschel, were "spacious, elaborate and habitable worlds." The stars were "effulgent centres of life and light to myriads of unseen worlds."[35] Herschel's arguments in favor of this conviction were classic examples of British natural theological argument. "Now, for what purpose are we to suppose such magnificent bodies scattered through the abyss of space?" he queried. "Useful, it is true, they are to man as points of exact and permanent reference; but he must have studied astronomy with little purpose, who can suppose man to be the only object of his Creator's care, or who does not see in the vast and wonderful apparatus around us his provision for other races of animated beings."[36] John was not the only defender of the elder Herschel's reputation. W. H. Smyth, admiral in the British navy and astronomical enthusiast, asserted that the "inhabitants of every world will be formed of the material suited to that world, and also for that world, and it matters little whether they are six inches high, as in Lilliput . . . whether they crawl like beetles, or leap fifty yards high."[37] These were theological arguments—assertions that a recognition of the plurality of worlds was also a recognition of

[35] J. Herschel, *Treatise on Astronomy* (London, 1833), 2.
[36] Ibid., 380.
[37] W. H. Smyth, *Cycle of Celestial Objects* (London, 1844), 1: 92.

God's power and benevolence. The popular writer Jane Marcet in her best-selling *Conversations on Natural Philosophy* (1819) was an advocate of pluralism as well.

Not all British commentators concurred with this assessment. An equally best-selling popular writer, Mary Somerville, argued in her *Connexion of the Physical Sciences* (1834) that "the planets, though kindred with the earth in motion and structure, are totally unfit for the habitation of such a being as man."[38] Sir Charles Lyell deployed his geological expertise to cast doubt on the prospect of "the plurality of habitable worlds throughout space, however favourite a subject of conjecture and speculation."[39] Lyell's friend Charles Darwin, on the other hand, was a fan of pluralism, at least in his younger days. More dangerously for gentlemen of science, arguments in favor of the existence of extraterrestrial life lay at the center of the subversive *Vestiges of the Natural History of Creation*. The anonymous author (later revealed to be the publisher Robert Chambers) argued of the planets that "every one of these numberless globes is either a theatre of organic being, or in the way of becoming so . . . Where there is light there will be eyes, and these, in other spheres, will be the same in all respects as the eyes of tellurian animals, with only such differences as may be necessary to accord with minor peculiarities of condition and of situation."[40] Discussions of the possibility, at least, of the plurality of worlds was part of the common currency of astronomical debate in Britain as well as the German lands, and the balance of opinion seemed if anything to veer towards the positive.

In 1853, however, the English natural philosopher William Whewell delivered a devastating critique of the pluralist position in his anonymous pamphlet, *Essay on the Plurality of Worlds*. Whewell had previously been at least sympathetic to the possibility of extraterrestrial life, suggesting in 1833 that stars other than the Sun might also "have planets revolving about them; and these may, like our planet, be the seats of vegetable and animal and rational life."[41] By the 1850s, however, disgusted and shocked by the success of *Vestiges* and the impious uses to which it put the pluralist argument, the polymathic master of Trinity College, Cambridge—a devout Anglican and staunch Tory—had changed his mind. He now wanted

[38] M. Somerville, *Connexion of the Physical Sciences* (London, 1834), 264.

[39] Quoted in M. Crowe, *The Extraterrestrial Life Debate*, 223.

[40] [R. Chambers], *Vestiges of the Natural History of Creation* (London, 1844), 161–64.

[41] W. Whewell, *Astronomy and General Physics* (London, 1833), 207.

to show that "dim as the light is which science throws upon creation, it gives us reason to believe that the placing of man upon the earth (including his creation) was a supernatural event, an exception to the laws of nature. The Vestiges has, for one of its main doctrines, that even this was a natural event, the result of a law by which man grew out of a monkey."[42] By this reading, any defense of pluralism was in danger of descending into an argument in favor of natural law and a denial of special creation. Whewell argued strenuously that "in the eyes of any one who accepts the Christian faith," the Earth could never be "regarded as being on a level with any other domiciles. It is the Stage of the great Drama of God's Mercy and Man's Salvation." The "assertions of Astronomers when they tell us that it is only one among millions of similar habitations"[43] demanded strenuous refutation.

Whewell's diatribe certainly raised eyebrows. As one reviewer commented, "We scarcely expected that in the middle of the nineteenth century, a serious attempt would be made to restore the exploded idea of man's supremacy over all other creatures in the universe; and still less that such an attempt would have been made by one whose mind was stored with scientific truths. Nevertheless a champion has actually appeared, who boldly dares to combat against all the rational inhabitants of other spheres; and though as yet he wears his vizor down, his dominant bearing, and the peculiar dexterity and power with which he wields his arms, indicate that this knight-errant of nursery notions can be none other than the Master of Trinity College, Cambridge."[44] The venerable Scottish natural philosopher, Sir David Brewster, pitched in with venom, dismissing Whewell's arguments as degrading astronomy and subverting true religion. In the unkindest cut of all, he even compared Whewell (unfavorably) to the author of the heretical Vestiges. The Rev. Baden Powell, Oxford's Savilian Professor of Geometry, was more circumspect. His view was that "by the light of inductive analogy, astronomical presumption, taking the truths of geology into account, seems to be in favour of progressive order, advancing from the inorganic to the organic, and from the insensible to the intellectual and moral in all parts of the material world."[45] He was unambiguous that the "material world" included other

[42]Quoted in M. Crowe, The Extraterrestrial Life Debate, 275.

[43][W. Whewell], Essay on the Plurality of Worlds (London, 1853), 44–45.

[44]Quoted in M. Crowe, The Extraterrestrial Life Debate, 282.

[45]Baden Powell, Essays on the Spirit of Inductive Philosophy, the Unity of Worlds, and the Philosophy of Creation (London, 1855), 231.

planets orbiting around other suns as well. John Herschel's response in a letter to the besieged master of Trinity was dismissive: "So *this* then is the best of all possible worlds—the *ne plus ultra* between which and the 7th heaven there is nothing intermediate. Oh dear! Oh dear!"[46]

One of the most prolific and popular expositors of the argument for extraterrestrial life during the second half of the nineteenth century was Camille Flammarion. Born in Montigny-le-Roi in 1842, by the age of sixteen he had persuaded the eminent Leverrier, discoverer of Neptune and by then director of the Paris Observatory, to hire him as an apprentice. At the age of twenty, he published his first book, *La Pluralité des Mondes Habités*, an audacious and robust defence of pluralism that went through numerous editions (fifteen by 1870) and was translated into several languages. Unlike his patron Leverrier, who was a practitioner of Airy-style industrial astronomy, Flammarion was an enthusiast for astronomical physics and the possibilities it held of providing real information about the physical constitution of other worlds—and of their inhabitants. For Flammarion there was nothing about the Earth that marked it out as being particularly fit for life. There was nothing unique about humankind's habitation and everything to suggest that other worlds might prove at least as hospitable to life. Far from the Earth's being the world best established for the maintenance of life, a great number of other worlds were far superior in terms of inhabitability to our own humble planet.

If there really was life on other planets, it was a short step to ponder how communication might be established between the inhabitants of Earth and those of other worlds. Telegraphy and later telephony and wireless telegraphy established the possibility of communicating over vast distances. An increasing number of commentators from about the 1860s onwards speculated whether the vast interplanetary and interstellar chasms might be bridged in similar fashion. The Frenchman Charles Cros came up with the suggestion that electric light rays could be focused using parabolic mirrors so as to be strong enough to be detected by any inhabitants of Mars or Venus that might be looking at Earth through their telescopes. He proposed a code to establish communication. In 1891 Flammarion announced the bequest of 100,000 francs to the Académie des Sciences to establish a prize for the first person to communicate with the inhabitants of another planet and receive a reply within the next ten years. In England Francis Galton (Charles Darwin's cousin) suggested in the *Times* that signals from mirrors reflecting sunlight might

---

[46]Quoted in M. Crowe, *The Extraterrestrial Life Debate*, 311.

be detected by telescope-wielding Martians. In America the flamboyant inventor Nikola Tesla pronounced that "with an expenditure not exceeding two thousand horsepower, signals can be transmitted to a planet such as Mars with as much exactness and certitude as we now send messages by wire from New York to Philadelphia."[47]

One form of communication with life from other worlds was held to have already taken place. Analysis of meteorites—generally accepted to have an extraterrestrial origin—seemed to indicate that many contained carbon-based substances of organic origin. If these stones falling from the sky came from other worlds, then the organic remains they contained were clearly the remains of the indigenous life forms of those other worlds. The physicist Sir William Thomson, searching around for evidence to confute Darwinian evolution, quickly latched onto the possibilities. In his presidential address to the British Association for the Advancement of Science in 1871 he announced that "because we all confidently believe that there are at present, and have been from time immemorial, many worlds of life besides our own, we must regard as probable in the highest degree that there are countless seed-bearing meteoric stones moving about through space." Furthermore, the "hypothesis that life originated on this earth through moss-grown fragments of another world may seen wild and visionary; all I maintain is that it is not unscientific."[48] Evolutionists, sensing they were the target of Thomson's speculations, reacted scornfully. The prospect of life itself crossing interplanetary space was, however, very much in the air when H. G. Wells penned the *War of the Worlds* in 1898.

It was no coincidence either that Wells chose Mars as the subject of his fiction. The Planet of War hit the headlines in 1877 with the announcement by the Italian astronomer Giovanni Schiaparelli that he had discovered an extensive system of canals on the planet's surface. This was unambiguous proof that Mars was not only inhabited but inhabited by intelligent beings. Schiaparelli enjoyed a solid reputation as a cautious and reliable observer. He had studied with Encke in Berlin and Struve in Pulkowa before becoming director of the Brera Observatory in Milan in 1862. For more than two decades after his momentous discovery, astronomers across Europe and America lined up on one side or another of the disputed question: had Schiaparelli really seen canals or were they

[47] Quoted ibid., 398.

[48] W. Thomson, "Presidential Address," *Reports of the British Association for the Advancement of Science*, 1871, 41: 269–70.

7.4 The canals of Mars as observed by Percival Lowell in 1905.

an optical illusion? The popular historian of astronomy Agnes Clerke, writing in 1885, was in no doubt, however, that the canals' existence had been fully substantiated. In 1894 the American astronomer Percival Lowell joined the fray. At his Lowell Observatory in Flagstaff, Arizona, Lowell confirmed that he too had seen the canals and put forward the theory that they were designed to carry meltwater from the planet's polar icecaps to the equator. Disputes concerning the canals' reality (and the evidence they afforded of Martian life) carried on well into the twentieth century and were grist for the mills of a generation of science fiction writers (figure 7.4).

Debates concerning the possibility of extraterrestrial life caught the Victorian public imagination for a variety of reasons. Such discussions intersected with a number of major cultural concerns. Extraterrestrial life had important theological consequences. To some it was evidence of God's munificence and the reliability of natural theological arguments. Others like Whewell came to regard the plurality of worlds as suggesting a dangerous dilution of Anglican doctrine. To many radical advocates such as Robert Chambers the plurality of inhabited worlds was, like the nebular hypothesis, proof of the universal operations of natural law and hence of the possibilities of human social (and spiritual) progress. In William Thomson's hands it became an anti-Darwinian bludgeon. Astronomers— particularly those advocates of astronomical physics—picked up on arguments concerning extraterrestrial life as providing a powerful new incentive and affirmation of the cultural relevance of their labors. Their endeavors could through such debates have important things to say about mounting late nineteenth-century concerns about humanity's place in the Universe. Extraterrestrials represented both fears and hopes concerning what humankind's own future in the coming new century might be. As Schiaparelli speculated about the Martian society that had built the canals, he concluded that the "institution of a collective socialism ought indeed to result from a parallel community of interests and of universal solidarity among the citizens ... The interests of all are not distinguished from the other; the mathematical sciences, meteorology, physics, hydrography, and the art of construction are certainly developed to a high degree of perfection; international conflicts and wars are unknown; all the intellectual efforts which, among the insane inhabitants of a neighbouring world are consumed in mutually destroying each other, are unanimously directed against the common enemy, the difficulty which penurious nature opposes at each step."[49]

## Conclusion

Astronomy had never really corresponded to its romantic image as a solitary science in which lonely watchers scanned the skies from their watchtowers. By the end of the nineteenth century it corresponded to that image even less. Astronomy was a labor- and time-intensive exercise that demanded the allocation of resources on an impressive scale. Successful observatories needed the managerial regimes of factory production. From the beginning of the century, astronomers were already the recipient of considerable state patronage throughout Europe. It seemed a prerequisite of imperial expansion. Astronomers' careful cataloguing of the skies could lead to ways of more accurately positioning ships at sea—solving the problem of longitude. Such knowledge could put a seafaring nation anxious for overseas expansion and nervous of the territorial ambitions of its neighbors at a distinct advantage. As well as their apparent utility, astronomers rode high on Newton's reputation. His *Principia* was celebrated for having placed the science of celestial motion on an apparently certain footing. Astronomers could predict the future movements of the planets with clockwork confidence. Their science had gathered for itself a reputation for mathematical exactitude that was the envy of other

---

[49] Quoted in M. Crowe, *The Extraterrestrial Life Debate*, 515.

natural philosophers. It was the model discipline against which others measured themselves.

Nineteenth-century astronomers fashioned themselves and their science so that they appealed to a broad swathe of constituencies. Astronomy remained an important adjunct of the state. Precision about the stars could deliver precision about political geography as well. Observatories became factorylike centers of calculation, machinelike in their reliability. The science of the stars could be made to matter for terrestrial politics too. Understanding the history of stellar evolution delivered important messages about the proprieties of contemporary social organization. This turned telescopes into potential weapons of insurrection that merited careful policing. Astronomers such as Lord Rosse at Parsonstown needed to be careful what use was made of their discoveries. Not just anybody could be allowed to speculate on the meaning of the heavens. The anonymous author of the *Vestiges of the Natural History of Creation* was ridiculed by gentlemen of science for his presumption in opining on matters beyond his ken. Only specialist astronomers had the nous to properly interpret the message of the stars. At the same time, moreover, large parts of astronomy were becoming adjuncts of physics. Men such as William Huggins argued that their work had brought celestial phenomena literally into the physical laboratory and the lecture theater. In this respect, while astronomers continued to provide physicists with important lessons in the management of large-scale institutions, physics by the end of the nineteenth century had become the dominant partner. Physics rather than astronomy was *the* nineteenth-century science.

# 8

## *Places of Precision*

Precision measurement seems to us to be at the very heart of modern physics. Measuring their effects as accurately and precisely as possible seems a prerequisite for understanding how the laws of nature operate. This preoccupation is, however, a comparatively recent phenomenon. It was only during the nineteenth century that laboratory disciplines started to put a whole new emphasis on precision measurement. Particularly towards the end of the century, as many physicists concluded that the end of physics was nigh—that they had established the general laws by which the Universe operated—the task at hand seemed to be one of consolidation. With few fundamental discoveries left to make, measurement seemed to many the path to a scientific reputation. Finding more and more ingenious ways of determining the exact value of constants and units was the big task ahead for physics. Establishing common standards of measurement was seen as the key to progress. For physicists such as James Clerk Maxwell, this was a moral crusade as well. Maxwell argued that the absolute identity of molecules was proof positive that they were manufactured articles fresh from some celestial production line. If they were manufactured articles then they required a designer—God. On this view there was a direct line between the physicist's routines of precision measurement and Victorian Anglicanism. Maxwell's remarks provide another clue as well to account for this concern with precision. Laboratories by the

226

second half of the century were increasingly part of an industrial culture that depended on disciplined regimes of accuracy and exactitude. Factories depended on finely measured, identical, and interchangeable components just as laboratory physics depended on reliable, robust, and universal constants.

Laboratories in the eighteenth century were few and far between. A few individuals—those who could afford to do so—maintained private laboratories in their own homes, where they carried out their own researches. Universities, however, did not maintain laboratories in anything resembling the modern sense. There might be a room or an annex behind a lecture theater where demonstrations were prepared, but these were not places of research—and neither were university professors expected as part of their duties to carry out any such research. Laboratories proliferated during the nineteenth century, however. No longer spaces of private exploration, they increasingly became centers of research and—just as importantly—teaching. Physics professors were expected to pass on their experimental skills as much as their book knowledge to the next generation. In Britain, France, Germany, and North America, institutions of higher learning jostled to acquire a physics laboratory—and preferably an eminent physicist to direct it. Laboratories became part of the trappings of a modern university. Students would learn the skills of precision measurement in a carefully disciplined and regulated atmosphere. Even as far afield as Japan, European experimental physicists were imported to establish teaching laboratories and pass on the increasingly vital skills of accurate experimentation.

In Britain by the end of the century, the preeminent leader of the pack was without doubt Cambridge's prestigious Cavendish Laboratory. By the early years of the twentieth century the majority of physicists manning university physics laboratories throughout Britain and its colonies had passed through its portals. The Cavendish manufactured experimental physicists as surely and successfully as it manufactured reliable measurements. The Cavendish's success was not achieved without encountering opposition, however. Many worried that a laboratory might not sit well with the reputation of an ancient university catering to the needs of the sons of the upper classes. Some of the dons were certainly concerned that there was more than a whiff of the factory floor about a late Victorian laboratory—hardly an appropriate adornment then for a civilized institution. James Clerk Maxwell, as the first Cavendish Professor, had to work hard to convince them otherwise. He had to find ways of integrating the laboratory into the university's established, hallowed regime.

The Cavendish's reputation by the end of the century was proof of his success. He and his successors, Lord Rayleigh and J. J. Thomson, had transformed the place from a potential thorn in Cambridge's side to a real rose in the university's crown.

For Cantabrigian physicists and other Britons, the major competition during the final decades of the nineteenth century seemed to be coming from the Germans. For much of the century German physics seemed to be in the ascendancy, having taken over from an early lead by the French under Laplace's leadership. British physicists certainly pointed to German physics as being at the root of the new German state's rising industrial (and military) clout and lobbied for increased government funding accordingly. The Germans themselves, however, were less confident. They saw their own physics institutions and laboratories as being in just as much dire need of reform. In particular, according to the electrical entrepreneur and industrialist Werner von Siemens, the new Reich needed its own physical laboratory to keep the opposition at bay, and he was prepared to put up the money for it. The result was the Physikalisch-Technische Reichsanstalt, founded in 1887 after more than a decade's planning and with the eminent Hermann von Helmholtz at its head. The aim was to create an institution devoted to the imperial, industrial, and intellectual needs of the ambitious new state. The fledgling institution would compete with and outstrip the best in Europe in the production of scientific standards, making physics a tool of German industrial progress and expansion.

The most ambitious—and consequential—of the late nineteenth century's grand standardizing projects was the scheme to found an international system of electrical units. Standards like these were deemed essential for the burgeoning international telegraph cable industry on which European and American imperial expansion increasingly depended. As well as ruling the waves, the late nineteenth century's colonial powers needed to have fast and reliable ways of communicating with their distant peripheries. This meant a network of underwater telegraph cables criss-crossing the globe. That network's reliability depended crucially on electrical standards. To maintain the highest efficiency, the cables' electrical characteristics—particularly electrical resistance—had to be known with precision. This was what underlay the British Association for the Advancement of Science's campaign to establish a reliable unit of resistance—the ohm—in the 1860s. Early efforts were spearheaded by Maxwell at King's College London and the apparatus moved with him to Cambridge in the 1870s. Under Rayleigh, standardizing the ohm became a major focus of the Cavendish's activities. It was no accident that British researchers led the field here—most of the world's undersea telegraph network was British owned. One of the Physikalisch-Technische Reichsanstalt's ambitions was to muscle in on Cambridge's preeminence in this area.

By the end of the century, large physics research and teaching laboratories were a part of the cultural landscape. Across Europe, North America, and beyond, universities without such facilities were rapidly becoming the exception rather than the rule. Such places were widely recognized as being powerhouses of industrial culture. As well as producing the hosts of experimental physicists needed to fill new university positions, these laboratories produced disciplined cadres of engineers and technicians destined for careers in industrial laboratories and workshops. These institutions were wedded to a cult of precision. Making better and ever more accurate measurements of nature's constants was the order of the day. The emphasis on precision fostered discipline. That was the key to unlocking nature's secrets. Increasingly as well, this laboratory discipline was coming to be regarded as a saleable commodity. By the end of the century physicists had a recognizable career structure stretching from undergraduate training through supervised postgraduate research to an industrial or academic position. Spokesmen for the discipline argued for ever larger allocations of public funds to expand the profession. Physics, they argued, had a crucial role to play in fin de siècle culture and in furthering social and economic progress. It was a key weapon in any industrial nation's armory.

## The Rise of the Laboratory

Until well into the nineteenth century, institutional laboratories for research—and more crucially, for teaching—in physics were something of a rarity. In no European country was research, in anything resembling the modern sense of the word, taken to be part of the duties of a university professor, for example. A professor's role was regarded as pedagogical—his task was to transmit established knowledge to his students, not to produce new knowledge. University lecture theaters might have an annex—typically behind the lecturer's podium—where demonstrations were prepared, but there was no clear distinction between the backroom work of experiment and the front-of-house activity of demonstration. In some ways, the emergence of the laboratory as a distinctive research—and pedagogical—space in its own right can be thought of as the building of a wall between those two spaces. Research was coming to be regarded as an autonomous activity in its own right rather than as an adjunct to teaching. As such it was seen as requiring its own institutional spaces. Furthermore, it was coming to be regarded as something for which a specific regime of training was needed as well. By the middle of the nineteenth century, laboratories were, if not ubiquitous yet, certainly more common as institutional spaces. The structures of French academies were revamped and research started to be recognized as part of a Faculty member's remit; German physics professors insisted that their cabinets of philosophical instruments be overhauled; in Britain, natural philosophers pointed to Continental developments and insisted on the need to emulate them.

French experimental natural philosophy and its institutions enjoyed a high reputation already at the beginning of the nineteenth century. British and German natural philosophers regarded with envy the facilities and state support their French counterparts were offered. French physical sciences, like the rest of science, had been systematically reordered following the Revolution and under Napoleon's imperial rule. The result was a strictly regulated and hierarchical system of institutes and faculties, largely revolving around Paris. French universities were reorganized under Napoleon into a single University of France, with faculties in the various provincial centers, including science faculties. Experimental and mathematical physics were high on the agenda, though subservient to medicine and law. Foreign students anxious to imbibe the best possible natural philosophical education flocked to Paris to study at the École Polytechnique and the Sorbonne. The laboratories where they clamored to train with masters such as Gay-Lussac and Jean-Baptiste Dumas were not state financed, however. They were private domains. French laboratory physics had a history of concern with precision stretching back to Coulomb's and Lavoisier's pioneering experiments at the end of the eighteenth century. It was an integral part of the Laplacian approach to physics and survived the demise of the Laplacian program.

By the second half of the century, however, French physicists were increasingly agitated by what they saw as the decline of their science. French physics institutions seemed moribund compared to the innovations taking place at German universities, or even in Britain. Foreign students seemed more interested in studying in Berlin than in Paris. French politicians and industrialists worried as well about the way French industry seemed to lag behind its European competitors. The message was rammed home by France's disastrous military defeat by Prussia in 1871. French

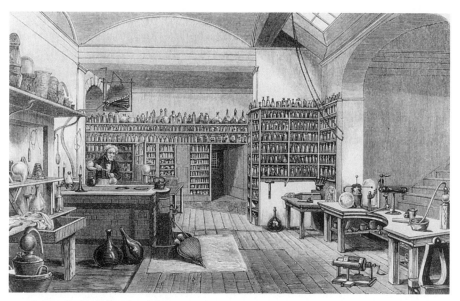

8.1 Michael Faraday cheerfully at work in the basement laboratory of the Royal Institution.

physics advertised itself as a solution to the problem. Technical education and laboratory training could produce new cadres of proficient engineers and technicians who would revolutionize French industry. Science faculties quite deliberately sought to transform themselves into institutions with a direct industrial role. As much as preparing teachers for the nation's *lycées*, scientific directors now regarded themselves as committed to the task of producing a scientifically literate workforce. Auguste Lamy at Lille in the 1850s, for example, tailored his research and his teaching on thermodynamics to fit the needs of local Lillois industry, drawing in crowds in the process. Later in the century, science faculties across France aimed to produce engineers for the nation's burgeoning electrical industry. As in Britain and Germany, the inculcation of precision physics in the laboratory was seen as having a distinct payoff in national productivity.

In Britain, one of the first institutional (as opposed to private) laboratories with more than just a supporting role for its adjacent lecture theater was the one at the Royal Institution. Sir Humphry Davy and following him Michael Faraday used the institution's laboratories to further their own research activities (figure 8.1). Originally a chemistry laboratory, with Faraday in charge the work done in the Royal Institution's basements shifted towards physics experiments. The laboratory was used for Faraday's own private research and not to train students (he never had any) in laboratory skills and disciplines. As university professors acquired more institutional space for their experimental activities, however, it became more common for some of them to encourage favored students to join them in their laboratories. Nowhere was experimental training a formal prerequisite of academic study. Enthusiastic favorites could learn the rudiments of experiment at their masters' feet instead. James D. Forbes, professor of natural philosophy at the University of Edinburgh, for example, encouraged students to spend time with him in the laboratory. The status of these laboratories was often ambiguous. The physical space might be provided by the university, but more often than not it was the individual professor who provided the apparatus out of his own pocket, as did Charles Wheatstone at King's College London, with the experiments that led to his invention with William Fothergill Cooke of the electromagnetic telegraph.

When William Thomson arrived back in Glasgow in 1846 to take up his position as professor of natural philosophy, he was anxious to take advantage of the opportunity to embark on his own ambitious program of experimental work. He soon turned to his students to provide him with assistance in this labor-intensive business and gradually, as "other students, hearing that some of their class fellows had got experimental work to do, came to me and volunteered to assist in the investigation,"[1] the basis of an academic teaching laboratory began to form. Arrangements were formalized in the mid-1850s with the university agreeing to provide Thomson with laboratory space that could accommodate his students as well. By 1862, Thomson could boast to Helmholtz that "I have had a really convenient and sufficient laboratory for students. Out of about 90 who attend my lectures, about 30 have applied for admission to the laboratory, and of these 20 or 25 will work fairly. I hope I may have half a dozen who will do good work."[2] By then they were helping Thomson with his work on electrical measuring apparatus, with an eye to the telegraph industry and their mentor's role in the plan to lay a telegraph cable across the Atlantic. They were learning the value (in all senses of the word) of precision measurement.

Thomson's Glasgow innovations established a model that other British physicists sought to emulate. In 1866, George Carey Foster established

---

[1] Quoted in G. Gooday, "Precision Measurement and the Genesis of Physics Teaching Laboratories in Victorian Britain," 31.

[2] Quoted ibid., 35.

a physics teaching laboratory at University College London. He was followed later the same year by Robert Clifton at Oxford. In 1868 Peter Guthrie Tait formalized James Forbes's old arrangements and established a teaching laboratory at the University of Edinburgh. By 1885, speaking at the opening of the physics laboratory at University College Bangor in North Wales, William Thomson could claim that "[t]he physical laboratory system has now become quite universal. No University can now live unless it has a well-equipped laboratory."[3] Others were still a little less gung-ho on the issue. Reminiscing about his appointment at Liverpool in 1881, Oliver Lodge recalled that "it was no joke having to start a laboratory from the beginnings and collect all the apparatus. Physical laboratories were rather novelties in those days. Carey Foster's had been the first of its kind in England; I mean a place where students were trained to perform experiments for themselves."[4] At Liverpool, Lodge found himself in the unenviable position of equipping a physics department in the rooms of an old lunatic asylum. The padded room, he noted, became incorporated into his laboratory. According to Lodge, British models were still so few and far between even by the 1880s, that he felt obliged to "make a tour of the Continental laboratories and gain experience that way."[5]

For Britain's new breed of laboratory managers, the apparently inexorable rise of the laboratory was proof of their field's progress and evidence of a new, disciplined ethos of precision. Frederick Guthrie, professor of physics at the Royal School of Mines, argued in 1870 that "the exact and experimental sciences are now so much more fully developed that it is impossible to remain any longer contented with attempting to teach an experimental class by means of a blackboard and a piece of chalk; it is now necessary to have an efficient apparatus for teaching these subjects."[6] Robert Clifton at Oxford concurred. As he sought to convince the university in 1868 of the need for new facilities, he insisted that "it has now become necessary for students to achieve fuller instructions than can possibly be given in public lectures and it is as important for a student of physics to become acquainted by actual experience with accurate

[3] Quoted ibid., 42.

[4] O. Lodge, *Past Years* (London: Stodder & Haughton, 1931), 153.

[5] Ibid.

[6] Quoted in G. Gooday, "Precision Measurement and the Genesis of Physics Teaching Laboratories in Victorian Britain," 36.

physical processes, as it is for students of chemistry or physiology to receive practical instructions in these departments of science."[7] The point
was to inculcate an ethos of precision. This was what progress in physics
depended upon.

Britain was not alone in fostering the cult of precision and identifying
the laboratory as the key to progress in physics. As Oliver Lodge's remarks indicate, some Britons did indeed worry that they had been rather
slow in getting the message. Many looked to the German example for
inspiration. Germans themselves were proud of their successes. Werner
von Siemens boasted in 1883 that "[n]o nation in the world has done
so much for scientific and technical education as Germany, and especially Prussia."[8] As in Britain, university teaching laboratories were very
much a nineteenth-century innovation. From the early eighteenth century, German natural philosophy professors had habitually assembled
physical cabinets—collections of instruments usually for demonstration purposes—which sometimes extended to lecture theaters and workshops. Again as in Britain these were usually the property and responsibility of the individual rather than the institution. In 1833, however,
Wilhelm Weber established a laboratory at the University of Göttingen
where he offered his students hands-on experience of performing experiments. He continued the practice and established a laboratory at Leipzig
when he moved there in 1837. He was emulated by Heinrich Gustav
Magnus in Berlin in 1843 and by Franz Neumann at Königsberg in 1847.
Both these were private initiatives in the professors' own homes. It was
not until 1862 that Magnus persuaded the university to take over the
laboratory's funding.

German states for much of the first half of the century took the view
that scientific discovery was not the business of the university instructor.
It was not an attitude that encouraged universities to provide laboratories.
By the 1860s, however, state-funded university laboratories were becoming more common. Crucially, the teaching regime increasingly included
a *Praktikum*: a course of training in laboratory experimentation. By the
1870s as well, more and more German universities were establishing their
own physics institutes. The expansion of experimental physics—and the
cultivation as in Britain of the cult of precision—was helped by German
physicists' insistence that new German industries, particularly the rising
electrical industry, were "the children of physics," for whom the time

[7] Quoted ibid., 38.
[8] D. Cahan, "The Institutional Revolution in German Physics," 1.

had come "to reimburse their Mother for her former nursing."[9] As Emil Warburg argued in 1881, if physics were the source of Germany's new-found industrial might, then new institutes and laboratories were essential to meet "the daily increasing need of successfully communicating in a contemporary manner the theories and methods of physics to a larger audience and of winning a larger number of arms for the advancement of science."[10] Men of physics such as Rudolf Virchow argued that the impact of physics on national life would not just be "the ever-greater extension of material productivity" but to make it "the maxim of our thinking and of moral action."[11] Physics could teach the virtues of precision in everyday life as well as in the laboratory. The bible of the new German physics institutes was Friedrich Kohlrausch's *Leitfaden der praktischen Physik* (1870) with its lengthy discussions of techniques of precision measurement and exhaustive tables of constants.

Increasingly throughout the century, American natural philosophers and physicists were working to convince their institutions of the importance of laboratory work in the production of scientifically literate and proficient students. When the electrician Joseph Henry was hired from Albany by New Jersey College in Princeton, he was adamant that he should have a laboratory equipped with the best European instruments. He set out on a trip to London and Paris to stock up on the essential apparatus. The laboratory he so stocked was for his private use and for the preparation of lecture demonstrations only, however. His students' contacts with experiment were limited to Henry's own spectacular performances in the classroom. There was no hands-on experience. By the end of the century, however, American physicists had fully imbibed their European counterparts' expressions of enthusiasm for hands-on laboratory training and the importance of precision. Many had acquired their own training in experiment at the feet of European masters in Britain and Germany and brought the cult of precision back home with them. Henry Rowland at Baltimore's Johns Hopkins University was a particular enthusiast. The spectroscope diffraction gratings he developed at his Hopkins laboratory set new standards in precision that impressed even the Europeans. A colleague noted on a trip to Europe in the 1880s that the "Germans spread their palms, looked as if they wished they had ventral fins and tails to express their sentiments"[12] concerning Rowland's inventions.

---

[9]Ibid., 39.   [10]Ibid., 41.   [11]Ibid., 40.

[12]Quoted in G. Sweetnam, "Precision Implemented," 284.

By the 1870s the newly restored Meiji regime in Japan was looking with interest at the successes of Europe's precision laboratories. Meiji officials were anxious to transform imperial Japan into an industrial power comparable with, if not superior to, European and American economies. To this end they were keen to recruit British physicists in particular to staff new scientific institutions and pass on the tricks of the experimenter's trade. The Imperial College of Engineering in Tokyo, established in 1873, hired William Ayrton—one of Thomson's Glasgow stalwarts and a veteran of the Indian Telegraph Department—to introduce Japanese students to the delights of electrical engineering. By 1877 the students under Ayrton's command were "well practiced in the construction and use of galvanometers, electrometers, resistance coils and condensers, etc., and in the performance of all the tests employed in a land line, or submarine cable testing office: artificial lines having been arranged, as far as practicable, with resistance coils, condensers, and a hundred yards or so of the Atlantic cable that were at our disposal."[13] By the time Ayrton returned to London in 1878 to put the teaching skills he had acquired in Tokyo to work at the London City and Guilds Institute, his Japanese students had received a grounding in the business of precision measurement that was the envy of any Western laboratory. His star student, Rinzaburo Shida, graduated to work with William Thomson in his Glasgow laboratory before returning to Tokyo in 1883 as professor of telegraphy at the Imperial College of Engineering.

Precision transformed laboratories across Europe and beyond into industrial powerhouses. The "spirit of accuracy" was lauded as not only providing a much needed guarantor of present success and future progress in physics, but as underpinning the future usefulness of physics as well. Like Victorian factories, physics laboratories could be depended upon to produce a steady stream of diligently designed, mass-produced, and standardized products. It was a public expression and validation of the self-discipline that careful training inculcated into the science and its practitioners. Being precise was both difficult and easy. The culture produced carefully standardized instruments and units that were seemingly straightforwardly transferable from one place to another. On the other hand the business of precision experiment required real commitment and arduous training. Producing an accurate measurement of a physical phenomenon that would pass muster with an increasingly hard-nosed

[13] Quoted in Y. Takahashi, "William Ewart Ayrton at the Imperial College of Engineering in Tokyo," 200.

and competitive community of expert physicists required a great deal of hard labor. Despite (or perhaps because of) the finely balanced precision instrumentation that increasingly filled these new academic laboratories, painstakingly acquired skills were the order of the day. Just as we saw earlier that grueling preparation underlay the apparently effortless analytical performances of Cambridge mathematicians, hard work was essential to make the grade as an experimenter too.

## The Making of the Cavendish

By the end of the nineteenth century, any survey of physics laboratories would certainly have identified Cambridge's Cavendish Laboratory as one of the most successful both in terms of teaching and research (figure 8.2). In the course of little more than a quarter century, Cambridge had established itself—to all appearances securely—as being among the foremost producers of experimenters and experimental physics. The first three Cavendish Professors—James Clerk Maxwell, Lord Rayleigh, and J. J. Thomson—remain stellar figures in the firmament of physics. To many eyes at the end of the century, Cambridge seemed synonymous with experimental physics. Achieving that identification was not an easy task, however. Establishing a role for laboratory physics and the cult of precision in that venerable institution was by no means straightforward. Dons used to dealing with the sons of gentlemen—and wedded to the ideal of a liberal education—needed considerable persuasion that an establishment of a sort more usually associated with grubby industry than pure intellect was really worth having. Its promoters had to persuade the doubters that laboratory physics, despite the unpromising appearance, was a fit vehicle for the promotion of liberal educational ideals—that it could train and discipline the mind appropriately. It was a dilemma of which the Cavendish's early professors were keenly aware and worked hard to overcome.

In 1869—and in the teeth of considerable opposition—a university committee reported to the senate on the vexed question of the teaching of experimental physics at Cambridge. They noted that the "importance of cultivating a knowledge of the great branches of Experimental Physics in the University was prominently urged by the Royal Commissioners appointed in 1850 to inquire into the State, Discipline, Studies and Revenues of the University and Colleges of Cambridge."[14] Topics such as

---

[14] Quoted in J. G. Crowther, *The Cavendish Laboratory, 1874—1974*, 23.

8.2 An exterior view of the Cavendish Laboratory on Free School Lane, Cambridge.

electricity, heat, and magnetism were occupying an increasingly important place in the examinations for the mathematics Tripos, and it seemed
a good time to consider how those subjects might be better taught. It was
no coincidence that another royal commission—presided over by the
duke of Devonshire, the university's chancellor—was soon to be considering the question of scientific and technical education throughout the
country. The committee concluded that it was time for the university to
establish a professorship of experimental physics and that the "founding
of a Professorship would be incomplete unless means were also supplied
to render the Professor's teaching practical, and assistance given to him,

8.3 A group of laboratory students in the Cavendish Laboratory. Note the presence of a
female student.

both in the Laboratory and Lecture-room. The need of providing instruments, Apparatus and Laboratories is obvious, and it seems not less
necessary to obtain some additional assistance in giving personal instruction to students in the Laboratory"[15] (figure 8.3). While the university
dithered over finding the cash, the duke of Devonshire announced his
willingness to fund the proposal out of his own considerable pocket.

William Cavendish, the seventh duke of Devonshire, had himself
been second wrangler and Smith's prizeman at Cambridge. He was also
a successful and prominent industrial entrepreneur. His offer of largesse
was well timed. It broke the colleges' resistance to the proposal, and
the search was soon on for Cambridge's first professor of experimental
physics. The obvious candidate was William Thomson, by then easily
the most eminent British physicist. He was unwilling to leave Glasgow,
however. The electors then looked to Hermann von Helmholtz in Berlin,
but he too was unwilling to abandon the security of his position there

[15] Quoted ibid., 24.

for an uncertain future at Cambridge. The eventual choice was James Clerk Maxwell, at the time retired to live the life of a country laird on his Scottish estates at Glenlair. Maxwell was initially dubious. He had, as he said, "no experience of this kind." Furthermore, he worried that the "Class of Physical Investigations, which might be undertaken with the help of men of Cambridge education, and which would be creditable to the University, demand, in general, a considerable amount of dull labour which may or may not be attractive to the pupils."[16] He was eventually persuaded, however, and took up the chair in 1871.

Maxwell was well aware from the outset that his fragile position would require careful consolidation if he were to overcome the dons' suspicions. The new laboratory, when it opened in 1874, must not look too much like a workshop or, as Maxwell worried, "we may bring the whole university and all the parents about our ears."[17] In his inaugural lecture he was anxious to reassure his auditors that the new laboratory under his direction would be a center for far more than mere mechanical drudge work. Precision measurement mattered indeed but should not be confused with industrial production. If that were all it was, then Maxwell argued, "Our Laboratory may perhaps become celebrated as a place of conscientious labour and consummate skill, but it will be out of place in the University, and ought rather to be classed with the other great workshops of our country."[18] This was a paradox that needed careful management, since after all, precision measurement was to be at the core of the Cavendish Laboratory's activities. Maxwell presented it as something that fitted eminently well with the university's natural theological tradition. Showing the high standards of accuracy and precision with which nature had been mass-produced was, argued Maxwell, a good way of demonstrating the powers of its designer: "we may learn that those aspirations after accuracy in measurement . . . are ours because they are essential constituents of the image of Him who in the beginning created, not only the heaven and the earth, but the materials of which heaven and earth consist."[19]

Maxwell visited Thomson at Glasgow and Robert Clifton at Oxford searching for guidance on how to design and organize his new laboratory. He hired William Garnett, fourth wrangler in 1873, as demonstrator. The laboratory was stocked with apparatus (paid for by the duke of

[16] Quoted ibid., 34.

[17] Quoted in S. Schaffer, "Late Victorian Metrology and Its Instrumentation," 33.

[18] Quoted ibid., 25.

[19] Quoted in S. Schaffer, "Metrology, Metrication and Victorian Values," 460.

Devonshire) mostly pertaining to the physics deemed relevant to the mathematics Tripos such as heat, electricity, and magnetism. Students were few and far between. There was no requirement that undergraduates studying for the mathematics Tripos, or even the natural sciences Tripos established in 1854, should attend at the laboratory. Maxwell was keen that those students that there were, like W. M. Hicks, later professor of physics and vice-chancellor at Sheffield University, worked under their own steam. He set Hicks to work on measurements of electrical resistance and of the Earth's magnetic field. Students were encouraged to make their own apparatus, which according to Hicks was "worth any amount of routine measurement."[20] Maxwell's scheme was to give his students a thorough grounding in measurement and the use of instruments— overseen by Garnett as demonstrator—before they embarked on their own projects. The British Association's Kew magnetometer, transferred to Cambridge, was used to introduce students to the vagaries of experiment and the skills they needed to acquire. Each student entering the laboratory was, like Hicks, given the task of using the magnetometer to measure the Earth's magnetic field.

Maxwell's tragic death at an early age in 1879 left the Cavendish looking for a successor who could continue the legacy of precision experiment that he had bequeathed it. The favored candidate was Lord Rayleigh, another aristocratic former wrangler with a considerable scientific reputation. Rayleigh was initially unwilling, worrying that such a position might be beneath his dignity, despite encouragement from the duke of Devonshire. William Thomson pleaded with him that if "you could see your way to take the Chair it would I am sure be much to the benefit of the university, and of science too, as the Cavendish Laboratory would give you means of experimenting and zealous and duly instructed assistants and volunteers and would naturally lead you to more of experimental research than might be your lot, even with all your experimental zeal and capacity for investigation, if you remain independent"[21] (figure 8.4). What clinched it for Rayleigh, however, was the agricultural depression, which made it impossible for him to maintain his private laboratory at his country estate in Terling, Essex, in the style to which he had become accustomed. He deigned to take the professorship for a period of five years, by which time to hoped the depression might be over. Despite

[20] Quoted in J. G. Crowther, *The Cavendish Laboratory, 1874—1974*, 62.

[21] R. J. Strutt, *Life of Lord Rayleigh* (London: Edward Arnold & Co., 1924), 100.

8.4 The Cavendish Laboratory's lab assistants in 1900.

his initial reluctance, however, Rayleigh arrived in Cambridge with some very definite ideas concerning the Cavendish's future direction.

Rayleigh wanted a massive expansion of the Cavendish's teaching program; in particular he wanted to increase the number of undergraduates trained in the laboratory—most of Maxwell's students had already graduated when they studied there. The new regime was to be systematic, modeled on Helmholtz's practice in Berlin. Under the new dispensation "each experiment was set out permanently on a table to itself, and written directions were provided. The classes were at regular hours, and a demonstrator was in attendance, who assigned the experiment, and gave help in any difficulty, finally approving or disapproving the numerical result."[22] Rayleigh hired two new demonstrators, R. T. Glazebrook and W. N. Shaw, to replace Garnett. Their textbook, *Practical Physics*, based on the Cavendish course, soon became a classic of laboratory physics instruction. In April 1880, students were informed that the "Cavendish Laboratory will open daily from 12 April for the use of students, from

[22] Ibid., 105.

11 am until 5 pm, and the Professor or Demonstrators will attend daily to give instruction in *Practical Physics*. The fee for the use of the Laboratory will be two guineas per term."[23] Training began from the basics—Rayleigh complained that at the beginning "[a]nyone who could handle a thing without knocking it off the table was an acquisition"[24]—and aimed to give a thorough grounding in the discipline of precision measurement. As well as teaching them experimental skills, Rayleigh was inculcating his students with laboratory discipline. They were being taught the value of systematic and patient routines of inquiry.

Rayleigh also wanted to put the Cavendish's research on a systematic footing. He was emphatic in his own experimental work concerning the importance of systematization. Glazebrook remarked of his superior's researches that they were "marked by the same characteristics: perfect clearness and lucidity, a firm grasp on the essentials of the problem and a neglect of the unimportant. The apparatus throughout was rough and ready, except where nicety of workmanship or skill in construction was needed to obtain the result; but the methods of the experiments, the possible sources of error, and the conditions necessary to success were thought out in advance and every precaution taken to secure a high accuracy and a definite result."[25] These were the values to be inculcated into the Cavendish's ethos. As Sir Arthur Schuster recalled, however, Rayleigh's concern with system went beyond his individual researches: "One idea to which he attached importance and which was entirely his own, was to identify the laboratory with some research planned on an extensive scale so that a common interest might unite a number of men sharing in the work."[26] Rather than being a place for the demonstration of individual experimental virtuosity, Rayleigh planned the Cavendish as a center for collaborative enterprise. The enterprise he chose was an ambitious one. He wanted to establish the Cavendish as a center for the redetermination of electrical standards—a project with which Maxwell had been involved from the 1860s. Such a project if successful would establish the laboratory firmly at the heart of physics.

Five years after his appointment—and with the agricultural depression comfortably behind him—Rayleigh resigned the Cavendish Professorship and returned to his private laboratory at Terling. His successor

---

[23] Quoted in J. G. Crowther, *The Cavendish Laboratory, 1874–1974*, 88.

[24] R. J. Strutt, *Life of Lord Rayleigh* (London: Edward Arnold & Co., 1924), 106.

[25] Quoted in J. G. Crowther, *The Cavendish Laboratory, 1874–1974*, 98.

[26] R. J. Strutt, *Life of Lord Rayleigh* (London: Edward Arnold & Co., 1924), 109.

at the Cavendish, J. J. Thomson, was already familiar with the value system he left behind, having worked at the laboratory himself under Rayleigh's direction. From a respectable, middle-class Mancunian background, Thomson had studied at Manchester's Owens College before gaining a scholarship to Trinity College, Cambridge. He crammed with the wrangler-making coach E. J. Routh to graduate second wrangler and gain a coveted Trinity College fellowship. It was only then that he started working at the Cavendish Laboratory, a month after Maxwell's death. The aim was to acquire for himself a thorough grounding in the rudiments of precision experiment to go with the grounding in mathematical physics with which the Tripos had equipped him. This was very much in line with Maxwell's own vision of the laboratory, as J. J. Thomson put it, as "a place to which men who had taken the Mathematical Tripos could come, and, after a short training in making accurate measurements, begin a piece of original research."[27] When appointed to replace Rayleigh in 1884, Thomson was young but already a fellow of the Royal Society. His appointment surprised many—including Rayleigh—who still thought of Thomson as more of a mathematician than an experimentalist.

The work Thomson had undertaken at the Cavendish under Rayleigh's direction had been a classic piece of Maxwellian experimental physics. His challenge was to determine the ratio of the electrostatic to the electromagnetic units of electric charge. According to Maxwell's theory the ratio should be equal to the velocity of light. The project had the potential, therefore, to be a bravura demonstration of the program of systematic precision measurement that Rayleigh was putting in place at the Cavendish as well as a powerful vindication of Maxwell's theory of electromagnetism. Thomson's efforts were not, however, particularly successful, and he soon moved on to help Rayleigh with his grand collaborative project to redetermine electrical standards. He was cutting his experimental teeth on cutting-edge technology. By the mid-1880s his reputation as a diligent and productive experimenter was high. He was a good example of what the Cavendish regime under Rayleigh could produce. Moving on to find his own experimental projects as professor, he took up the study of cathode rays of the kind pioneered, as we saw earlier, by William Crookes. The research would pay dividends with his discovery of the electron a decade or so later.

J. J. Thomson's mounting reputation as an experimenter underscored the Cavendish's own rising star. He continued and consolidated

[27]J. J. Thomson, *Recollections and Reflections* (London, 1936), 95.

Rayleigh's program of systematic experimental instruction. As he recalled, the "number of science students in Cambridge increased rapidly after 1885."[28] A new wing to the Cavendish was opened in 1896 to relieve the pressures of massively increasing numbers. There was "a very large room used for the elementary classes in practical physics, for examinations in practical physics for the Natural Science Tripos and for entrance scholarships to the Colleges. Besides this there was a new lecture-room, cellars for experiments requiring a constant temperature, and a private room for the Professor."[29] In 1895 it became possible for graduates of universities other than Cambridge to enter the Cavendish as research students, initially for the degree of M.A. but later gaining Ph.D.s. As Thomson noted, "since the M.A. degree did not entitle a man to be called 'doctor', our students were at a disadvantage when competing for teaching posts with those who had been to a German university and had obtained the Ph.D. degree."[30] The influx (and eventual outflux) of foreign students allowed by the new regulations enabled the spread of the Cavendish's values of precision—and its reputation—across Europe, America, and the Empire.

By the end of the nineteenth-century Cambridge and the Cavendish stood at the heart of an expanding worldwide network of competing laboratories and institutes. It also represented the core values of Victorian physics. Maxwell had regarded the laboratory as an outpost through which the ethos of precision measurement could be introduced to and intertwined with the Cambridge culture of liberal education. Exact measurement could be a way of broadening and disciplining the mind as well. Care and diplomacy had been required to reassure a skittish university establishment that bringing a laboratory to Cambridge was not after all tantamount to converting the college cloisters into factory floors. Maxwell's plan had been to make the place available to graduates who had already persuaded the university of their trustworthiness by going through the Tripos ritual. Under Rayleigh and J. J. Thomson the Cavendish did indeed become more of a production line, albeit one churning out only a small number of quality items every year. Its students were systematically introduced to a carefully worked-out culture of experimental discipline through carefully graded routines. The values they imbibed and exported as they scattered across the empire and the globe were ones of discipline, diligence, and precision as the hallmarks of experiment.

[28] Ibid., 123.    [29] Ibid.    [30] Ibid., 137.

## Berlin's Imperial Institute

As J. J. Thomson's concerns about his students' job prospects indicate, there was little doubt in British laboratory physicists' minds as to who their main rivals were. By the 1880s, Germany's new physics institutes looked like a formidable force. By the standards of envious onlookers they appeared well funded and well organized. To some German physicists, nevertheless, however enviable their institutions might seem compared with those in other European countries, things were still not good enough. Increasingly during the 1870s and 1880s they lobbied the new government in Berlin for more resources. In particular, they wanted an institution devoted to research not teaching. While British and French physicists pointed to the awesome reputation of German physics and its apparent role in boosting the new state's rapidly expanding industries, some of their German counterparts were insisting that far more needed to be done if future German industrial supremacy were to be achieved. They were anxious to persuade the Bismarckian state that physics had a key role to play in the Reich's future. The eventual result of this lobbying was the Physikalisch-Technische Reichsanstalt, set up in 1887 with Hermann von Helmholtz as its director, committed to physics research as an arm of the imperial state. The new institution was to be a powerful rival to the Cavendish in the world of precision measurement.

The resources that some German states committed to their physics institutes were already formidable. During the 1870s, Prussia committed more than 1.5 million marks to Helmholtz's Berlin Institute. The British physicist John Tyndall, Faraday's successor at the Royal Institution, remarked enviously to a colleague that "you will find in the Berlin laboratory the very things which my American and British friends and I should like to see in operation in all college and university laboratories in America and in the British Empire."[31] The powerful and influential industrialist Werner von Siemens argued, however, that this was simply not enough. His complaint was that "[s]cientific research itself, however, is not a professional activity within the state structure; it is only a tolerated private activity of scientists alongside their profession . . . The sad consequence of all this is that in most cases scientific projects that might revive and stimulate entire areas of life are not undertaken."[32] The German states encouraged and financed physics teaching, according to

---

[31] Quoted in D. Cahan, "The Institutional Revolution in German Physics," 23.

[32] Quoted ibid., 1.

Siemens, at the expense of progress in research. The institutes produced legions of teachers instead of the experimenters who could make a real contribution to the Reich. The solution in his view was a new imperial institution committed to research.

Werner von Siemens was a powerful voice in the new Germany. Siemens and his brother had made their fortunes as pioneers in the new electrical industries that emerged during the second half of the nineteenth century. While Werner took care of the German end of affairs, his brother Wilhelm emigrated to Britain, eventually taking up British nationality and turning himself into Sir William Siemens. During the 1840s, Werner von Siemens, along with his partner J. G. Halske, was at the forefront of German telegraphy. He regarded himself as a physicist as much as an industrialist. "My love always belonged to science as such," he said, "while my work and accomplishments lay mostly in the field of technology."[33] In 1884, Siemens decided to put his money where his mouth was and offered to fund the establishment of a Reich physics institute. He had made a similar offer to Prussia the previous year. It was to be an institute for research alone: "The teachers and laboratories of the universities and pedagogical schools are not appropriate for the purpose; neither are the professors employed by them. The more active these latter are and the more they have proved themselves to be pathbreaking researchers, the more they are overburdened by their teaching obligations and the extra duties bound up with them." He was convinced that "[f]rom the planned natural scientific workplace, both material and ideal advantages of great importance would accrue to the Reich."[34]

Lobbying in favor of some kind of national or imperial physics research institute had been going on since the early 1870s. Men such as the physiologist Emil du Bois Reymond, Wilhelm Foerster (director of the Berlin Observatory), and Hermann von Helmholtz argued that Prussia and the Reich needed some kind of institute devoted to precision measurement. It was an argument that found favor with Prussian military strategists such as Helmut von Moltke. Opinions were mixed as to what exactly the new institute should do. Helmholtz wanted a body that granted funds for precision instruments. Others wanted a commercial testing station. Little concrete happened until the 1880s and Siemens's offer. Siemens argued that "England, France, and America, those countries which are our most dangerous enemies in the struggle for survival,

[33] D. Cahan, *An Institute for an Empire*, 36.
[34] Ibid., 40–41.

8.5 An artist's sketch of the proposed new Physikalisch-Technische Reichsanstalt.

have recognized the great meaning of scientific superiority for material interests and have zealously striven to improve natural scientific education through pedagogical improvements and to create institutions that promote scientific progress."[35] Despite Siemens's offer, persuading the Reich's bureaucracy—particularly gaining Bismarck's indispensable support—was difficult. Some powerful interest groups, including engineers, industrialists, and physicists, worried that the proposed Reichsanstalt would encroach onto their own territory. It was not until 1887 that eventual agreement was secured.

The proposed institution needed careful planning. It would need "well-planned rooms protected from external disturbances, excellent and costly instruments" as well as "the complete devotion of the scientists." Siemens was worried that "Bismarck . . . still holds science for a type of sport without practical meaning" and that a great deal of work still needed to be done to convince him otherwise. The public and fellow physicists needed to be convinced that this would be "a place of work open to all outstanding German scientists"[36] and not just for a cabal of Berlin insiders (figure 8.5). In Siemens's plan, the institute would be divided into two

[35] D. Cahan, "Werner Siemens and the Origins of the Physikalisch-Technische Reichsanstalt," 204–5.

[36] Ibid., 276.

sections—physical and technical—under the control of an overall president. The technical section would be responsible for choosing scientific problems, setting the budget, and generally administering the institute. The physical section would have as its main task the development of new experimental investigations. The technical section was to be subdivided into five carefully chosen subsections, representing areas where the Reich hoped for industrial supremacy: materials testing, precision mechanics, optics, thermometry, and electrical standards testing. Siemens had already chosen the man who would be in charge of this great new enterprise. His choice was the grand old man of German physics—Hermann von Helmholtz.

Helmholtz was widely recognized in Germany and elsewhere as being head and shoulders above his contemporaries. To one fan he was the "Imperial Chancellor of German Science." An American student studying under him remarked that "the whole scientific world of Germany, nay, the whole intellectual world of Germany, stood in awe when the name of Excellenz von Helmholtz was pronounced. Next to Bismarck and the old Emperor he was at that time the most illustrious man in the German Empire."[37] It was proof of his preeminence that when he was appointed to head the new Berlin Physics Institute in 1871 he could command the staggering sum of 315,000 marks to be spent on his official residence there. He had enough clout that he virtually held the Reich to ransom before agreeing to take up the position as the Physikalisch-Technische Reichsanstalt's president. He demanded a salary of 15,000 marks along with an annual bonus of 9,000. The government in the end had little choice but to capitulate. As they recognized, the institution had to a large extent been designed with Helmholtz in mind as its eventual director. Helmholtz enjoyed a wide reputation as well as a public spokesman for science in Germany. He seemed ideally suited for the task of putting physics in its proper place at the heart of the German state.

Helmholtz proved to be an inspirational leader at the Reichsanstalt, as he had at the Berlin Physics Institute. Like his counterparts at the Cavendish Laboratory, Helmholtz was keen to get his people working as a team. Once the institute's building was complete, the scientific section had its own *Observatorium* built for the purpose, designed to be free from external disturbances. The entire building was constructed on a thick thousand-square-meter concrete slab for maximum stability and the external walls were shielded from direct sunlight to help ensure a

---

[37] D. Cahan, *An Institute for an Empire*, 65.

constant temperature. Each of the floors was devoted to a different aspect of the Reichsanstalt's research. Thermodynamics work took place on the ground floor, where it was easiest to control the temperature. Electrical and optical work was on the highest floor, with the offices and library in the middle. The experimenters also had access to a machine house and a separate entirely iron-free building for magnetic experiments. These were unrivaled facilities that underlined the Reich's hopes for what physics could deliver.

Under Helmholtz's direction, the scientific section was divided into three laboratories working on heat, electricity, and optics. The heat laboratory focused on finding better materials for thermometers, improving the accuracy of thermometric measurements at high temperatures, and improving the design of heat engines—all precision projects. The electrical laboratory was in the business of competing with the Cavendish in providing accurate and reliable electrical standards—a matter of particular concern to Werner von Siemens—as well as experimenting on the effects of magnets. They carried out experiments for the Reich navy, trying to minimize the disruptive effects of iron on ships' compasses. At the optics laboratory, the main concern was to establish reliable industrial standards in the measurement of light, following Fraunhofer's achievements earlier in the century. This was a particularly pertinent concern as Germany led the world in optical instrumentation. Under Otto Lummer, Helmholtz's former student and assistant at the Berlin Physics Institute, the laboratory's workers experimented to develop a more reliable photometer—an instrument for comparing the intensity of light from different sources. The concern throughout was to establish standards of precision measurement that could be put to industrial use. Making such standards would be a tangible demonstration of German superiority in precision physics and a warning shot across the bows of its industrial competitors in the rest of Europe and America.

Disaster struck the Reichsanstalt in the *schwarze Jahr* of 1894. Helmholtz died. Finding a replacement was not to be easy. His deputy, Ernst Hagen, worried that it was "unforeseeable how the situation here at the Reichsanstalt will develop since Helmholtz has died . . . The main problem lies in the fact that basically everything here was tailor-made for Helmholtz's *person*."[38] The eventual choice as successor was Friedrich Kohlrausch, author of the ubiquitous *Leitfaden*. Beyond his reputation

[38] Ibid., 123.

as the author of one of Germany's most widely used physics textbooks, Kohlrausch had much to recommend him. He was a veteran administrator, having been in charge of five physics institutes before arriving at the Reichsanstalt. His father, Rudolf, had himself been an eminent experimentalist in midcentury and had ensured a fine training and good contacts with the best in the field for his son. Friedrich had acquired his doctorate with Wilhelm Weber at Göttingen in 1863 before going on to codirect the Göttingen Physics Institute with his former mentor later in the decade. When he received the call to Berlin, Kohlrausch was director of the Strassburg Physics Institute—one of the largest (and most expensive to build) in Germany.

As well as being an old hand at administration, Kohlrausch had something else to recommend him. He had built his career as an experimenter on the activity that was in many ways the Physikalisch-Technische Reichsanstalt's raison d'être—precision measurement. Looking back over his career in 1900, he opined that "measuring nature is one of the characteristic activities of our age."[39] From that perspective, Kohlrausch's activities had certainly been preeminently characteristic. A colleague, Heinrich Rubens, remarked that "no other physicist has surpassed Kohlrausch in the skill and care with which he used instruments and methods."[40] He had been responsible for inventing and developing a whole range of revolutionary new precision instruments—dynamometers, galvanometers, magnetometers, and reflectometers. In particular, he had made his reputation in the field of electrical measurement and the establishment of electrical standards. From the beginnings of his career working with Weber he had devoted almost forty years to working at determining the values of electrical and magnetic constants and units. He had represented German interests at many of the international congresses devoted to working out acceptable international standards of electrical measurement and was widely recognized as *the* German expert in that industrially crucial field. He was regarded as eminently well-placed to steer the Reichsanstalt towards helping ensure German domination of the expanding electrical industries.

The Reichsanstalt expanded massively under Kohlrausch's direction. It had more or less doubled in size by 1903. Kohlrausch devoted the same kind of diligence to his administrative tasks as he did to his vocation of

[39] Ibid., 129.
[40] Ibid., 130.

precision measurement. A former colleague, Svante Arrhenius, described him as having "always lived as orderly as a chronometer and is in all social relations a strict formalist. His principal scientific endeavour is directed at improving measuring methods, so as to make the probable error smaller. Indeed he has an all-too-great predilection for finely rounded numbers."[41] In Phileas Fogg style, Kohlrausch clearly expected those around him to conform to his orderly expectations. The informal regime inaugurated by Helmholtz as director was replaced by a far more formal and rigid administration. Kohlrausch agreed, however, with Helmholtz and the institute's founders about the Reichsanstalt's wider purpose. It was there to place physics at the service of the Reich. The different laboratories of the science section pursued much the same activities—albeit on a larger scale—as they had under the previous administration. Much of the institute's expansion took place in the technical section, which by the late 1890s was fully devoted to serving the needs of German industry for scientific testing.

The Physikalisch-Technische Reichsanstalt was in many ways an entirely unprecedented institution. In no other European country had the community of physicists persuaded their national government to support their research in such lavish fashion. It was testament to the success with which physicists had maneuvered themselves into positions of real power and influence in the new German Reich. Men of science such as Hermann von Helmholtz could wield political clout that was the envy of contemporaries elsewhere in Europe. His public pronouncements on the state of science and of its relevance to German cultural life mattered. The Reichsanstalt was celebrated at home and recognized abroad as a triumphant expression of an ambitious new industrial power's potential. The German physics community had been fashioned into a seemingly indispensable arm of the state. That fashioning had taken place around the cult of precision. As in Britain, physicists had successfully argued that their concern with precision measurement not only contributed to industrial progress, but that it also expressed important cultural values. The Physikalisch-Technische Reichsanstalt was an instantiation not only of the utility that the German Reich hoped for from the systematic application of physics to industry but of the virtues of self-discipline and application that the new state wanted to foster in its citizens.

[41] Ibid., 134–35.

## A Real, Purchaseable Tangible Object

The importance of international standards in physics for national industry (and for national prestige) was underlined in the ongoing battles that dogged the second half of the century surrounding the measurement and use of electrical standards. The debates capture not only the ways in which precision mattered for the formation of physics as a discipline and as an important, self-defining part of the physicist's art, but the ways in which precision expressed views concerning the cultural place of physics and its potential utility. Being in a position to define international standards in an increasingly important field of research like electricity put the victorious party in a position of some advantage. It meant that everyone else had to come to them to have their apparatus validated. It also signaled the consolidation of electricity as a science. As William Thomson pointed out, turning an electrical unit into "a real, purchaseable tangible object" so that "we may perhaps buy a microfarad or a megafarad of electricity,"[42] would bring physics fairly and squarely into the Victorian marketplace. It demonstrated that physics had value in a way even the most hardheaded industrialist could understand and appreciate. This was what standardization was all about. It encompassed the progress in scientific knowledge and in industrial supremacy that disciplined physics could deliver.

The spread of commercial electric telegraph networks from the 1840s onwards encouraged the creation of a new breed of electrical experts. Putting these networks together and—just as importantly—maintaining them once they were up called for extensive electrical know-how. Telegraph engineers such as Latimer Clark soon realized that one of their biggest problems was finding the location of faults in the lines. Particularly with underground cables, unless the engineer could find a way of more or less precisely locating a fault—such as a break—valuable time was lost and valuable labor wasted digging along the line to find the problem. Latimer Clark had joined the Electric Telegraph Company as an engineer in 1850, when the company was still easily the largest in Britain, from a background in civil engineering. He rapidly became a pioneer in the new field. Along with others such as Cromwell Varley, Clark realized that the key to finding faults in underground wires was measurement. A good knowledge of the characteristics of their copper wires—particularly

[42]Quoted in S. Schaffer, "Late Victorian Metrology and Its Instrumentation," 32.

of their resistance to the passage of current—was a prerequisite to finding faults. One simple method, for example, was simply to see what length of wire in the workshop gave the same galvanometer reading as the faulty wire. The length of the wire then determined the position of the break in the line. This only worked, of course, if the test wire and the cable had similar resistances. Ways were needed of calibrating the components of telegraph circuits.

The problem of measurement became more urgent from the 1850s onwards with the development of underwater telegraphy. The first commercial underwater cable was laid between Dover and Calais in 1851. It soon became clear that underwater telegraphy had its own particular problems. The problem of locating faults accurately and quickly became more urgent. After all, dredging for a faulty cable underwater was a considerably more costly business even than digging for one on land. Submarine cables also suffered from a problem known as retardation— signals tended to become smeared and merged into each other over long distances. The cables seemed to leak, so that more current was needed to ensure a good signal at the other end. All of this meant that telegraph engineers needed a good understanding of the electrical characteristics of their equipment. They needed to be able to measure those characteristics and they needed to be able to compare them effectively with their workshop or laboratory apparatus. By the late 1850s, telegraph engineers were therefore increasingly working with standardized pieces of equipment like resistance coils or condensers. As William Thomson recalled, looking back at the history of precision electrical measurement, "Resistance coils and ohms, and standard condensers and microfarads, had been for ten years familiar to the electricians of the submarine-cable factories and testing-stations, before anything that could be called electric measurement had come to be regularly practised in almost any of the scientific laboratories of the world."[43]

Matters came to a head in many ways with the ambitious plans of the late 1850s and early 1860s to lay down a telegraph cable across the Atlantic, linking the Old World with the New. The Atlantic cable's promoter, Cyrus Field, had coaxed a fortune from his backers to finance the enterprise, as well as securing the cooperation of British and American governments. It was a disaster when the first cable failed in September 1858 after barely a month of operation. Since 1857, William Thomson, a director of the Atlantic Telegraph Company, had been carrying out

---

[43]W. Thomson, *Popular Lectures and Addresses* (London, 1891), 1: 82–83.

experiments at his Glasgow laboratory on the cable being used in the enterprise, finding out in the process that it was of extremely variable quality. While Thomson argued that this was a major defect—albeit one about which little could be done since much of the cable had already been manufactured and was indeed in the process of being laid—the company's electrician, Wildman Whitehouse, argued that cable resistance (or conductivity) mattered little. The key to successful signaling in his view was the use of his patent induction coil apparatus to send rapid, high-intensity bursts of electricity down the cable. The pinpointing of Whitehouse's high-intensity jolts as one of the primary causes for the cable's failure did much to concentrate minds on Thomson's suggestion that strict quality control of cable production was essential.

Following the cable's failure, the British government and the Atlantic Telegraph Company convened a committee to oversee an inquest into its early demise. Taking evidence from a raft of telegraphic experts, they also commissioned research on an unprecedented scale into the electrical characteristics of telegraph cables. Latimer Clark alone carried out experiments on hundreds of miles of copper wire and the new insulating material gutta-percha. Another key witness was Fleeming Jenkin, a young engineer from R. S. Newall's cable factory at Birkenhead near Liverpool. He carried out extensive experiments on the relative resistances of copper and gutta-percha in an effort to calculate the amount of current that would leak out through the insulation in a telegraph cable. In comparing the resistances of different substances like this Jenkin in particular came up against the problem of standards. There were no commonly agreed units of electrical measurements in which he could express his results. There were by this time a number of resistance standards in use. Telegraph engineers in Britain used coils calibrated in miles of copper wire, in France they used kilometers of iron wire, and so on. What the new tests highlighted, however, was the very unreliability of these standards themselves. Their reliability depended on the purity of their components, which was exactly what the inquest to the failed Atlantic cable cast doubt upon.

Into this breach stepped the British Association for the Advancement of Science. At its Manchester meeting in 1861 it established a committee on electrical standards to look into the whole vexed question. The committee, which included Fleeming Jenkin, William Thomson, and Charles Wheatstone (inventor of the telegraph) in its ranks, was soon joined by James Clerk Maxwell. Their aim was to act on the suggestion, put to the British Association by Sir Charles Bright and Latimer Clark, that "the science of Electricity and the art of Telegraphy have both now arrived at a stage of progress at which it is necessary that universally received standards of electrical quantities and resistances should be adopted, in order that precise language and measurement may take the place of the empirical rules and ideas now generally prevalent."[44] It was not a straightforward matter. The committee needed to set up a standard that met the needs of laboratory physicists and practical telegraph men. The British Association's ohm, as the crucial standard of electrical resistance came to be called, was the product of much hard work and negotiation. Maxwell was a critical figure; it was at his laboratory at King's College London that the crucial experiments to establish the value of the ohm were carried out. When he received the call to Cambridge and the Cavendish Laboratory, he was determined that the British Association's ohm and its instrumentation would follow him there. It was too important a piece of intellectual and commercial property to leave behind.

In his groundbreaking *Treatise on Electricity and Magnetism*, Maxwell argued that "the determination of electrical resistance may be considered as the cardinal operation in electricity, in the same way that the determination of weight is the cardinal operation in chemistry."[45] Carrying out this cardinal operation was to be one of the Cavendish Laboratory's chief tasks. The British Association measuring apparatus consisted of a rapidly rotating coil with a magnetized needle at the center (figure 8.6). As the coil rotated in the Earth's magnetic field, a current would be induced that caused the needle to deflect. The size of the deflection depended on the diameter of the coil, the rate at which it was spun, and the coil's resistance. Measuring the needle's deflection, the coil's diameter, and the rate at which it was spun gave a highly accurate value for the coil's resistance. This would make it possible to produce a standard resistance coil, defined as one ohm. Simple as it might appear in principle, getting the experiment right required the mobilization of major resources. Some of Britain's most skilled engineers and scientists devoted themselves to the problem. Measurements of unprecedented precision were needed to get at the right level of accuracy.

Following Maxwell's death, his successor, Lord Rayleigh, made the project his own. Measuring the ohm would provide the Cavendish with a collaborative project that would help bind its workers into a disciplined, unified corps and establish a set of laboratory values in more senses than

---

[44]Quoted in B. Hunt, "The Ohm Is Where the Art Is," 58.

[45]J. C. Maxwell, *Treatise on Electricity and Magnetism* (Cambridge, 1873), 1: 465.

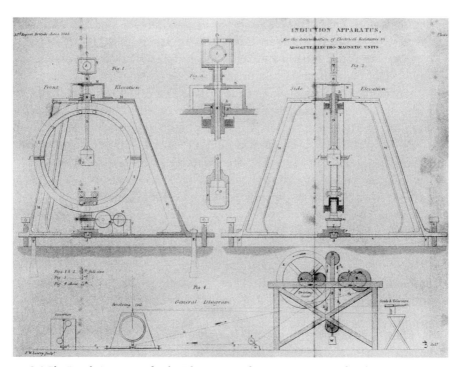

8.6 The British Association for the Advancement of Science's Committee for Electrical
Standards' apparatus for measuring the ohm.

one. As Rayleigh started work on the project in 1880, it was a concerted
team effort: "The apparatus had been set up on the ground floor of the
laboratory, in the room then known as the 'magnetic room' . . . The revolving coil was set up on a brick pillar . . . The observations were made late
at night, to avoid magnetic and other disturbance. Rayleigh regulated the
speed, Dr. Schuster took the main readings, and Mrs. Sidgwick recorded
the readings of the auxiliary magnetometer."[46] When Schuster left to
become professor of physics at Manchester, Eleanor Sidgwick—the wife
of the professor of moral philosophy and university reformer Henry
Sidgwick, and Lady Rayleigh's sister—took over his role, while Lady
Rayleigh herself often came in to replace her on the magnetometer. Triumphantly, the revised ohm gave a value for the mechanical equivalent of
heat, measured electrically, that tallied with the mechanically measured
value to an unprecedented degree of precision. It was a result that closely
tied the Cavendish ohm to the whole body of nineteenth-century energy

[46]R. J. Strutt, *Life of Lord Rayleigh* (London: Edward Arnold & Co., 1924), 114.

physics. James Prescott Joule hastened to congratulate Rayleigh on his success. "It is an extraordinary and gratifying result for all of us, and I congratulate your lordship and Schuster on the admirable experiments you have brought to so successful an issue,"[47] he wrote.

Rayleigh, the Cavendish, and the British Association ohm were not without their opponents, however. As early as 1851, Wilhelm Weber (Kohlrausch's old mentor) had published a highly sophisticated absolute system of standards based around units of force and motion. Following developments introduced by subsequent theorists, Weber's system had the advantage—to theoretically inclined physicists at least—of tying electrical quantities directly to the fundamental concepts of energy and work. Their definitions were highly complex, however. The resulting units were also very difficult to measure. Most seriously, Weber's units seemed of little use to jobbing electricians. Their values were too small to be of any practical use to anyone whose concern was to work with miles rather than inches of wire. The theory behind Weber's system was also increasingly suspect to British physicists weaned on Maxwell's field theories of electromagnetism.

More robust opposition came from Werner von Siemens, who advocated a standard resistance based on the use of columns of mercury. In his view, there was simply no need for any great metaphysical heart wrenching. All that was needed was to define the unit of electrical resistance in terms of the resistance of an arbitrary column of mercury. What mattered was that the unit chosen should be of a size useful for the telegraph industry: "those cases in which the expression of absolute measure is of advantage occur very seldom and only in purely scientific exercises," he argued. In Siemens's opinion, not only was the search for absolutism unnecessary, it was also suspect. It was a distraction from the task at hand. His standard (and arbitrary) mercury column resistances were handy simply because that was all they were designed to be—"every other definition would not only burden unnecessarily the calculations which occur in modern life, but also confuse our conception of the measure."[48] Part of the problem was that just as the British objected to Weber's system on the grounds that it involved adherence to Weber's theory, Siemens recognized that the British Association ohm, for

[47] Ibid., 117–18.

[48] W. von Siemens, "Suggestions for the Adoption of a Common Unit of Measurement of Electrical Resistance," *Reports of the British Association for the Advancement of Science*, 1862, **32**: 154.

example, increasingly embodied Maxwell's energy physics. Buying one meant buying into the other as well. Siemens's campaign to establish the Physikalisch-Technische Reichsanstalt throughout the 1870s and early 1880s was largely about trying to overcome this British imperialism in the field of electrical measurement and theory. The new institute could provide a rallying point for the opposition. At international congresses during the 1880s, the Germans fought hard against the British Association standard. It was a losing battle. In many ways, even the setting up of the Reichsanstalt was an admission of defeat. It was an acknowledgment that absolutism, not pragmatism, was the way to go after all.

The ohm demonstrated how physics furthered industrial, Victorian values. In many ways it is completely unsurprising that British electricians led the pack in the development of electrical standards and eventually dominated the field. After all, by the final quarter the century, British cable companies dominated the world's telegraph industry as well. The episode shows how the cult of precision fostered in physics laboratories across Europe fitted in with a wider set of values. Precision measurement could be seen as an answer to the question Victorian cynics often posed of physics—*Cui bono?* Whom does it benefit? What is it for? A great deal of labor was expended in the process. Establishing the ohm took mobilization on a grand scale. In Britain alone, laboratories at Cambridge, Glasgow, and London played central roles. Engineers, instrument makers, and physicists alike required and acquired new skills in the process. The biggest task of all was to make all of that labor appear invisible. All the customer purchased at the end of the day was a coil of wire. That coil stood for an absolute and universal system of measurement that was accepted as being independent of any local skills or resources. The whole point about the standard ohm was that despite the great efforts required to produce it in particular laboratories like the Cavendish, it was meant to work unproblematically anywhere in the world. In that way at least, the ohm was very much the epitome of Victorian values.

## Conclusion

From being rare, exotic places at the beginning of the nineteenth century, physics laboratories by the end of the century seemed ubiquitous. Every self-respecting university anywhere in Europe, the Americas, the colonies, or beyond needed a physics laboratory. As an institutional space, it had become part of the fundamental apparatus of learning. Not only elite universities, but even high schools might well have their own teaching laboratory by the end of the period. Such laboratories as existed in the previous century had as a rule been private places, the domain of particular individuals who had the resources to indulge in the experimental investigation of nature. By the end of the century most laboratories were public institutions. They were in principle open to all—all, that is, who had the appropriate credentials and qualifications. As the discipline of physics emerged out of natural philosophy during the course of the century, laboratories came to be that new discipline's archetypal institution, as experiment appeared to be the archetypal activity. They were the training grounds where acolytes acquired the skills they needed to investigate nature and put its products to work. Laboratories forged physics' links with industry and brought large parts of the industrial ethos with them into the citadels of academe.

Precision mattered for late nineteenth-century physics as a way of inculcating new disciplinary regimes as much as anything else. It was a crucial element in the fashioning of physicists as much as of physics. In the new academic teaching laboratories, the transmission of skills from mentor to student was a highly regimented process. Laboratory teaching, as well as what was taught, was increasingly standardized. The values of precision linked the world of late Victorian energy physics and its laboratories to the world of industry as well. Not only did the standardized electrical units produced in these laboratories have an immediate role to play in electrical industry, but the work regime and the ethos that had produced those units blended easily with those of industrial culture too. When Maxwell assured his audience of Cambridge dons that he had no intention of turning the Cavendish into a "manufactory of ohms," he was addressing a real concern. He was not entirely convincing in his denials either. Late Victorian laboratories manufactured physicists as well, moreover. At the beginning of the twentieth-century, Cambridge products were to be found reproducing the Cantabrigian ethos of precision all over the globe. The same could be said of Germany's physics institutes as well. Spreading the values of precision meant multiplying and disseminating the places of precision and its duly trained adepts too.

9

# Imperial Physics

On 17 December 1907, William Thomson, long since knighted and then ennobled as Baron Kelvin of Largs, died. His death symbolized the end of an era in European physics. He was buried next to Newton in Westminster Abbey. The ceremonial surrounding his funeral was a tangible demonstration not only of his own stature as a man of science but of the high place physics by then occupied in British and European culture. His first biographer, the physicist Silvanus P. Thompson, writing only a few years later remarked how the occasion "brought together one of the most wonderful congregations that has ever assembled in that historic building."[1] Representatives of innumerable British and foreign universities attended the proceedings. The lord mayor of London and the lord provost of Edinburgh were there. The duke of Argyll represented the king. Silvanus Thompson thought the pomp and circumstance indicated "some revival of recognition of what the nation owes to science and to her great men."[2] From his admittedly partial point of view he was confident that the "nineteenth century has, intellectually, been the golden age, not of art or of poetry, not of drama or of adventure, but of science."[3] Kelvin epitomized the values of "laborious humility" that physics stood for at the end of one century and the dawn of another.

[1] S. P. Thompson, *Life of Lord Kelvin* (London: Macmillan, 1910), 2: 1209.
[2] Ibid., 2: 1212.
[3] Ibid., 2: 1213.

261

Lord Kelvin's physics had indeed been the epitome of imperial science. The instruments he had invented—the highly sensitive mirror galvanometer in particular—and the theories he had promulgated underwrote Britain's supremacy in the field of international cable telegraphy. His work had been at the cutting edge of research that had played a key role in inventing this new discipline called physics. By the end of his life Kelvin in Britain—like Helmholtz in Germany—had come to stand for a particular kind of imperial physics. This was a cocksure new science, its spokesmen confident and self-assured, convinced that they held the keys to unlocking nature's secrets. To all appearances, late nineteenth-century (and early twentieth-century) physics was a spectacular success. It seemed uncontroversial to suppose that the general outlines at least of a comprehensive physical understanding of nature had been triumphantly established. Practitioners could tackle problems that would have been unheard of a generation or two earlier. They could analyze the makeup of the distant stars. They could send messages through the ether. They could measure the properties of ether and matter with hitherto undreamed-of precision. On the dark side, they could also predict the death of the Sun.

As we have seen throughout the previous chapters, none of this was achieved without a great deal of work and effort. The science of physics did not fall unaided from the laps of the gods. Neither did nature suddenly start speaking for itself. To make the kind of science that turn-of-the-century commentators and pundits marveled at and celebrated had required the marshalling of human and material resources on a previously unprecedented scale. Even if by twenty-first-century standards the numbers of practicing physicists and their laboratories at the beginning of the twentieth century were tiny, they represented a whole new order of magnitude when compared with previous centuries. To achieve this, physicists had to show that they could deliver the goods—they had to fashion themselves and their science so that they and it could be shown to matter for Victorian culture. This was not easily done. Many groups during the period regarded the whole idea of science with indifference, if not outright hostility. They had to be persuaded that this new science was, if not a positive benefit, then at least not a threat. Just as the practice of physics was molded to fit the contours of Victorian culture, that culture itself was remolded by the findings and practices of the newly dominant discipline.

Progress was the buzzword of the Victorian age. In civilization, in industry, and in the production of knowledge, the age seemed to be positively striding ahead. William Whewell smugly remarked that the Great Exhibition of 1851 provided a frozen tableau of human progress in the arts. At one end of the scale were the barbaric and primitive productions shipped in from outposts of empire. At the other were the sophisticated outpourings of American and European factory culture—entirely a "more skilful, powerful, comprehensive, and progressive form of art."[4] Different mid-nineteenth-century cultures represented different stages in the progress of mankind brought together as if in a time capsule at the Crystal Palace. Whewell, of course, argued as well that science had nothing to do with this. According to the master of Trinity College, progress in the sciences was something entirely different from progress in the mere arts of life. Radical natural philosophers earlier in the century argued that by demonstrating progress in the laws of nature they could establish progress as a necessary feature of social existence as well. The nebular hypothesis in their hands was an argument for social reform, if not for revolution. Later physicists demonstrated the progressive nature of their science in a different way. They reckoned they could show that physics was a prerequisite of the very industrial progress that so many of their fellow Victorians were so keen to celebrate.

Their physics reproduced that industrial culture as well. For physicists such as Kelvin or Maxwell, that their science forged a shared culture between the factory and the laboratory was one of its key features. The same discipline and the same attention to detail pertained in each. The same kinds of objects were to be found both in one and in the other. Physics ennobled manufacture by laying bare its affinities to the work of God. Molecules were evidence of design and a designer because they were all the same—just like the products of a Victorian engineering workshop. Precision measurement proved this and so could become a way of displaying the divine plan. This is probably the opposite of what we would think today. In an age like ours completely attuned to mass production, idiosyncrasy rather than uniformity seems a better index of intelligence and design. At the same time, affinities with the workshop exemplified physics' own utility. Campaigns to standardize and to tighten the boundaries of precision showed just what the disciplines of the laboratory could offer back to industry. The physics that governed the workings of the ether was just as relevant on the factory floor—and just as productive. Physicists placed themselves at the heart of late Victorian culture

[4]W. Whewell, "The General Bearing of the Great Exhibition on the Progress of Art and Science," *Lectures on the Results of the Great Exhibition of 1851* (London, 1852), 8.

by showing how they could harness nature's machinery in the name of material progress.

## Physics and Fin de Siècle Culture

Physics lay at the very heart of turn-of-the-century culture. Beyond its seeming promise to deliver hitherto unprecedented material progress and offer concrete solutions to the mysteries of the universe, it spoke powerfully to deeply held cultural concerns and fears. In new genres of popular literature physics was used to project the possibilities and dangers of future utopias and dystopias. The new science might be the vehicle for delivering Paradise on Earth. It might also be the key to building powerful new weapons that might destroy humanity. To a generation that regarded the prospect of a European war in the near future as an inevitability, some of the destructive possibilities offered by physics were disturbing indeed. Physics spoke to increasing concerns about the degeneration of the species as well. Turn of the century commentators viewed the prospects for humanity as bleak. Humans were degenerating under the impact of civilization. The middle classes were suffering from nervous strain and failing to procreate as they should. The physics of energy conservation as well as eugenics seemed to offer a solution as well as an explanation for the malaise. The same resource proved useful for men concerned with increasingly vociferous calls for women's education and political emancipation. Physics taught that women's bodies could not deal with the responsibilities of public as well as private life. Women's education would only accelerate the degenerative trend.

Two novels, *Frankenstein* and *Dracula*, provide interesting insights into deep changes that had taken place in public perceptions of physical science from one end of the century to the other. *Frankenstein*, written in the 1810s, in the century's infancy, dealt with fears concerning what natural philosophy might prove capable of delivering in the coming years. Its nightmare of artificial life unleashed betrayed fears of physics unrestrained. *Dracula*, at the other end of the century, raised the specter of physics' limitations. It was an English stiff upper lip and a dose of Yankee grit that saved the day at the end of the novel, not the scientific and technological paraphernalia that the protagonists turned to in their efforts to track the monster to his lair. Bram Stoker, *Dracula*'s author, was a graduate of Trinity College Dublin and well versed in the latest physics and its possibilities. The physicist who went too far, transgressing social and natural boundaries, was still a potent image though. Jules Verne's

Captain Nemo might be a brilliant scientist, but that did not save him from his submarine exile. If anything, the suggestion was that it was his single-minded pursuit of physics that had consigned him to the deep.

H. G. Wells's *Time Machine* (1895) postulated a bleak future indeed for humanity. In the *War of the Worlds* (1898) the nightmare was of humankind's impotence when faced with the threat of annihilation from a scientifically and technologically superior alien species. In the *Time Machine*, however, the enemies were all of humanity's own making. Thrust into the distant future by his machine, the Time Traveller found himself faced with Eloi and Morlocks, two separate species that were both degenerate descendants of humankind. Far from inaugurating an age of indefinite intellectual and material progress, science and technology had produced the opposite effect. Cosseted by their inventions and deprived of intellectual stimulation by their cocooned and protected lives, the middle classes had evolved into the childlike Eloi, living in a seeming Paradise but preyed upon by the bestial Morlocks, subterranean guardians of the machines their human ancestors had once operated. As the Time Traveller came to realize, the apparent utopia was just that: "The great triumph of Humanity I had dreamed of took a different shape in my mind. It had been no such triumph of moral education and general co-operation as I had imagined. Instead I saw a real aristocracy, armed with a perfected science and working to a logical conclusion the industrial system of the present day. Its triumph had not been simply a triumph over Nature, but a triumph over Nature and the fellow-man."[5]

Where Wells expressed a popular fear of the degenerative dangers of technically dependent civilization, a more optimistic gloss was offered by the prolific Edward Bulwer-Lytton in his only foray into what would now be called science fiction. The protagonist of *The Coming Race* (1871) found himself faced with a subterranean civilization compared to which surface-dwelling humans were at the stage of technological infancy. This underground utopia depended on the mysterious force, vril—"electricity, except that it comprehends in its manifold branches other forces of nature, to which, in our scientific nomenclature, differing names are assigned, such as magnetism, galvanism, &c."[6] Vril powered this mysterious civilization's machinery and gave its inhabitants seemingly magical mental powers, "akin to those ascribed to mesmerism, electrobiology,

[5] H. G. Wells, *The Time Machine* (1895; reprint, London: Everyman Library, 1995), 44–45.
[6] E. Bulwer-Lytton, *The Coming Race* (1871; reprint, Dover: Alan Sutton, 1995), 20.

odic force, &c., but applied scientifically through vril conductors."[7] Their control over this all-pervasive natural force had given the Vril-ya absolute dominance over their world. They controlled the weather and could wield weapons of unparalleled destructive capacity against any enemy rash enough to challenge them. Bulwer-Lytton's book was a massive hit, and vril entered the popular imagination. In one form at least it survives into the twenty-first century, despite its literary inventor's descent into comparative obscurity. Advertisers decided that bovine vril—or Bovril—would be the ideal name by which to sell the latest food product designed to improve the British diet.

Fears concerning racial and species degeneration were fueled by eugenic speculation about the likely fate of humankind if some of the debilitating effects of civilization were not overcome. Eugenic commentators pointed to the seemingly degraded bodily state of British army recruits for the Boer War as disturbing evidence of trends towards physical deterioration already in place. The conservation of energy provided a ready answer to the question of why such a decline was taking place. The human body contained only a finite amount of energy. Modern life with its frenetic pace took a heavy toll in nervous energy with the result that not enough was left to properly sustain the body's physical frame. Masturbation and other evils endemic in an increasingly effete and enervated culture only added to the problem. Men who squandered their precious reserves in solitary contemplation (of one sort or another) were not only endangering their own health, but the health of their children and the future of the race. For late Victorian and Edwardian commentators on their respective nations' health none of this was idle scaremongering. Their concerns, after all, were dictated by the latest theories in physics. Particularly in the early years of the twentieth century, when European prophets of doom reckoned that a Continent-wide war was practically inevitable, the social Darwinist consequences of failing to take seriously these threats to national virility seemed dire.

Physics offered a solution as well as an explanation of the new modern malaise, however. Where Bulwer-Lytton's fictional Vril-ya had vril, their real-life counterparts had electricity. Electricity had a long history as a panacea stretching back well into the eighteenth century. It had its therapeutic heyday, however, in the late Victorian and Edwardian age. It was widely touted as the latest scientific cure. Electric baths, belts, corsets, hairbrushes, pills, and rings were advertised as able to "give wonderful

[7] Ibid.

support and vitality to the internal organs of the body, improve the figure, prevent chills, impart new life and vigour to the debilitated constitution, stimulate organic action, promote circulation, assist digestion, and promptly renew the vital energy the loss of which is the first symptom of decay."[8] Electric gadgets might offer a discreet cure for the effects of those youthful indiscretions a respectable gentleman might be unwilling to share with his doctor. Electricity's perceived close relationship with the vital force had a downside as well. In 1890, New York carried out its first electrocution—a new term coined for the latest and most scientific method of judicial killing. Electricians quarreled with each other in the public prints as to whether this novel employment of their expertise was a stain or a bloom on their science's reputation.

If the conservation of energy gave men a hard time, women got an even rougher deal. Men were the victims of overcivilized society or of their own (correctable) vices. Women were at the mercy of their constitutions. Only so much energy could be contained in a woman's body. Its proper purpose was to be directed towards childbirth and nurturing, and woe betide the woman who diverted it to some other use. This was a law of nature rather than of society. As the psychologist Henry Maudsley pointed out in 1874, if "it were not that woman's organization and functions found their fitting home in a position different from, if not subordinate to, that of men, she would not so long have kept that position."[9] A woman's social place was an inevitable consequence of the laws of conservation: "it is not a mere question of larger or of smaller muscles, but of the energy and power of endurance of the nerve-force which drives the intellectual and muscular machinery; not a question of two bodies and minds that are in equal physical conditions, but of one body and mind capable of sustained and regular hard labour, and of another body and mind which for one quarter of each month during the best years of life is more or less sick and unfit for hard work."[10] The health and future of society dictated that women should conserve what meager energy resources they had for the rigors of childbirth.

This was an important argument for opponents of increasingly insistent women's voices calling for equal education and equal political rights. Women who overeducated themselves by this argument simply stopped being women. Overstimulation of the brain led to a redirection of the

---

[8]Quoted in I. R. Morus, "A Grand and Universal Panacea," 105.

[9]H. Maudsley, "Sex in Mind and Education," *Fortnightly Review*, 1874, **15**: 479.

[10]Ibid., 480.

vital nervous force away from the reproductive organs and towards the brain instead. The inevitable result was sterility, the loss of specifically female physical features such as breasts, and the development of male characteristics like facial hair. Any move to force women into ways of life that went against their nature as dictated by the laws of physics would be disastrous both for the individual woman who stood in danger of losing her essential femininity and for the human race as a whole. Commentators such as Joseph Mortimer Granville, inventor of nerve-vibration therapy, fulminated against "unwomanly women who lead their weaker but not less worthy sisters into these strange regions" as "the enemy of their sex."[11] He feared that the "strong-minded womanism of the day" would "sooner or later infallibly destroy the respect in which woman is held by man, and undermine the feeling which gives her a claim to his protection."[12] Ironically enough, just as physics was providing powerful ammunition for misogynistic diatribes such as these, women were increasingly demanding entry into physics laboratories for themselves.

This was not, therefore, by any means a unanimous opinion. Bulwer-Lytton endowed his female Vril-ya with physical and intellectual characteristics that were equal if not superior to those of the males. Several late nineteenth-century physicists welcomed women students into their laboratories and classrooms. They still had to overcome considerable resistance. While the patrician Lord Rayleigh went so far as to collaborate and even copublish with women physicists, his Cavendish successor J. J. Thomson took a rather dimmer view. He joked to a woman friend that "you would be amused if you were here now to see my lectures—in my elementary one I have got a front row entirely consisting of young women (some of them not so young neither, as someone says in Jeames' Diary) and they take notes in the most painstaking and praiseworthy fashion, but the most extraordinary thing is that I have got one at my advanced lecture. I am afraid she does not understand a word and my theory is that she is attending my lectures on the supposition that they are Divinity and she has not yet found out her mistake."[13] In his *Recollections and Reflections* Thomson reflected with equanimity on the violent opposition of Cambridge students to the admission of women to degrees. He was sure that while women's intellect might be up to the basics, they would be unable to cope with the complexities of advanced physics.

[11] J. Mortimer Granville, *While the "Boy" Waits* (London, 1879), 259.

[12] J. Mortimer Granville, *Youth: Its Care and Culture* (London, 1880), 98.

[13] R. J. Strutt, *Life of J. J. Thomson* (Cambridge: Cambridge University Press, 1942), 28–29.

Physics entered fin de siècle cultural discourse in many ways. It was a vehicle for the expression of deep fears about what a new century had to offer and about the limitations of material progress. Late Victorians were fascinated by the philologist Max Müller's ruminations on solar mythology. He read myths like the Nordic death of Balder (and maybe even the death of Christ) as metaphors expressing fears of solar destruction—that the Sun might not rise again in the morning. Victorian physics made those fears materially tangible with its predictions of the eventual death of the Sun, graphically visualized at the end of Wells's *Time Machine*. The conservation of energy proved to be a powerful tool with which to approach concerns about racial purity and danger. It could show how the stiff upper lip and self-discipline so beloved of late Victorian and Edwardian admirers of the imperial spirit were in fact prerequisites of human salvation. It provided good reasons for keeping women in their place. At a time when increasingly intense rivalry among nations was widely regarded as the inevitable order of the day, physics provided a diagnosis of the problem, a new arena for the staging of international competition—and a new source of potential tools for the job when war eventually came.

## Physicists at War

When European war did break out in 1914, physicists were quick to come forward, offering their science's services to their nation-states. There was nothing at all surprising about this. Physics had formed an arena for international competition for more than half a century. From their midcentury beginnings, international exhibitions of the arts and sciences had been regarded as showcases for industrial and scientific rivalries among nations. The British and French in particular looked on anxiously as the century progressed and succeeding exhibitions showed increasing evidence of growing German industrial might. European states celebrated their great men of science as symbols of national virility and power. Idealistic commentators might wax lyrical about the international nature of science, but national amour propre was in practice far more typical. Britons and Germans vied fiercely on the question of whether the conservation of energy was a British or a German discovery—though John Tyndall played the traitor and awarded the laurels to the German Robert Mayer. The British and French clashed over Adams's and Leverrier's claims to the discovery of Neptune. The potential dangers to national prosperity and security of increasing German scientific and industrial prowess was a constant

refrain from British and French physicists alike trying to coax more resources from recalcitrant governments throughout the final decades of
the nineteenth century.

As early as 1851 and the Great Exhibition at the Crystal Palace, British
commentators at any rate regarded the occasion as an opportunity to
show off the superiority of their own arts and sciences. In just the same
way they reacted with consternation to the Paris Universal Exposition
of 1867 and its evidences that national competitors were in the process
of stealing center stage. In Paris in 1900 some optimistic commentators
still saw exhibition culture as an antidote to internecine warfare: "The
Exhibition of 1900 had been, not only a success, but a benefaction. It
had served to relax the nerves of the French nation after a terrible drama
and it had brought about a truce, if not between parties, at least between
nations; hatred of the foreigner, so lively in 1899, was less virulent."[14]
Others, however, were more realistic in their assessment: "Economic
crowds merge in our Babel, along the rue des Nations, where each pavilion
upholds the ethnic character of its country, of its race. This contradiction
of a cosmopolitanism endorsed by all and a nationalism each day more
intransigent, everywhere more jealous of maintaining or restoring the
integrity of the race, of the mother-tongue, of the laws, the traditions: is
this not one of the great unknowns of the problematic legacy our century
leaves to its successor?"[15]

So where were the physicists in all of this? As usual, by the close
of the nineteenth century, they were at the center of things. Polemicists
such as Charles Babbage had been putting physics at the heart of exhibition culture since the 1850s. The exhibitions formed shop windows
for the latest physical discoveries, instruments, and inventions as well
as industrial applications. To exhibition-goers at least, demonstrations
of telephones, phonographs, or cinematography formed a seamless web
with displays of galvanometers, X-ray apparatus, or astronomical photographs. The industrial physicist Werner von Siemens was convinced
that exhibitions could "convince ourselves and the world that German
industrial hard work has developed to a higher level under the bounti-
ful rays of the recovered unity and power of the Reich, and that it can
engage the industrial competition of the world with calm assurance."[16]
And this was in the context of resurgent French triumphalism at the Paris

[14]Quoted in R. Brain, *Going to the Fair*, 152.

[15]Quoted ibid., 165.

[16]Quoted ibid., 163.

Exhibition of 1889, celebrating a hundred years of the French Revolution. Physicists of all nations played key roles in international exhibitions—as instigators, organizers, jurors of competitions. Such exhibitions formed useful venues for international scientific congresses of all kinds. They were perfect opportunities for showing the world how much physics mattered—and for reminding governments back home of the kudos that physics could attract and for showing off in front of the competition.

International scientific rivalry was particularly intense in the field of electricity—and with good reason. The massive underwater cable telegraph industry dominated by Britain was universally recognized as having immense strategic significance. Securing lines of telegraphic communication between the periphery and center of burgeoning empires was vital. Given British dominance of the industry things could be quite difficult for American, French, or German interests looking for lines of communication immune to potential, if not actual, British interception. Germany was to find out just how much this mattered in 1914, when Britain cut the cables between Berlin and the rest of the world. British strategists were just as keen to ensure that there was a secure "All-Red Route" safe from prying enemy eyes and ears. British governments paid handsome subsidies to telegraph companies operating alternative lines that sustained those vital All-Red connections. The rapid development of the new electrical power industries was adding to the science's significance as well. One perceptive American visitor to the Paris exhibition in 1900 (none other than Henry Adams) reported back that "[s]ince 1889, the great economy has evidently been electricity. Since 1840, electricity must have altogether altered economical conditions. Looking forward fifty years more, I should say that the superiority in electric energy is going to decide the next development of competition. That superiority depends, in its turn, on geography, geology and race-energy."[17]

A major battle in these internecine electrical wars took place in Paris in 1881. At that year's International Exhibition, representatives from various countries gathered together in an effort to solve the vexed issue of international electrical units. The British were anxious to defend the inroads into enemy territory made through the success of the British Association's ohm and the efforts of the Cavendish Laboratory's crack troops of precision measurers under Lord Rayleigh. The Germans were just as keen to recover lost ground and reassert the superiority of Werner von Siemens's mercury standard. They had sent their big guns to the

[17] Quoted ibid., 166.

conference table—Helmholtz, Siemens, and Emil du Bois Reymond. The British cause in turn was championed by the redoubtable Sir William Thomson. As Thomson's biographer recorded, "The debate grew warm. One who was present has narrated the unforgettable scene of comedy of Thomson and Helmholtz disputing hotly in French, which each pronounced *more suo*, to the edification of the representatives of other nationalities."[18] An uneasy truce was eventually established, with the British Association's measured value for the ohm being accepted by all combatants, but with the proviso that it should be concretely represented by a Siemens mercury column. Hostilities would be recommenced the following year over the question of just how long that mercury column ought to be. The French managed to win a minor skirmish as well, displacing the German Weber with their illustrious countryman Ampère as the name for the international standard of electrical current.

When battle resumed the following year, Thomson was desperate to marshal his forces. He wrote begging Rayleigh to attend at Paris. "We could get on but very badly without you," he complained, "and in fact I suppose your ohm must be declared the one and true ohm for our generation. Could you not, in an international cause of such importance, and in direct relations with your professorial work, arrange to have lectures postponed, and laboratory work cared for in your absence for the week or ten days which should suffice for Paris and the Committee?"[19] In the event, even though Rayleigh declined to drag himself away from the Cavendish, the British forces did well enough. Thomson reported back to the absent Rayleigh that the Cavendish squad's measurements were being accepted as definitive. Thomson dismissed discrepancies between the Cavendish's values for the Siemens unit and H. F. Weber's results with a sneer: "It seems after all probable that it was the particular one or two standards *called* Siemens units which we had that caused the discrepancy."[20] Desultory fighting over the issue continued for almost a decade. It was not until a conciliatory gathering at the British Association's Edinburgh meeting in 1891 that peace was finally declared with war-weary delegates from Britain, France, Germany, and the United States agreeing that the ohm was to be defined as the resistance of a column of mercury 106.3 cm long of 1 square mm area at a temperature of 0° centigrade—virtually no different in the end from the Cavendish value.

---

[18] S. P. Thompson, *Life of Lord Kelvin* (London: Macmillan, 1910), 2: 775.

[19] R. J. Strutt, *Life of Lord Rayleigh* (London: Edward Arnold & Co., 1924), 124.

[20] S. P. Thompson, *Life of Lord Kelvin* (London: Macmillan, 1910), 2: 790.

By the beginning of the twentieth century, then, these hard-fought-for, internationally agreed-on electrical standards were enshrined in the legal systems of various nation-states. In Britain, for example, electrical standards were policed and enforced by the government Board of Trade. If physics was part of the cultural fabric, it was coming to be an arm of the state as well. The possibility of setting up a national physical laboratory in Britain had been mooted at that very British Association meeting in Edinburgh in 1891 by Oliver Lodge. The gauntlet was taken up again halfway through the decade by the British Association, and a Treasury committee with Lord Rayleigh as chairman was set up to consider the matter. The result was the setting up of the National Physical Laboratory at Bushy Park—a former royal palace—under the supervision of the Royal Society. Rayleigh succeeded in having Richard Glazebrook, his onetime laboratory demonstrator at the Cavendish, appointed its director. The NPL inherited from the Cavendish the mantle of maintaining Britain's lead in the business of electrical measurement. Just as Germany's Physikalisch-Technische Reichsanstalt had the task of maintaining the Reich's increasingly formidable reputation in physics and its applications to industrial advancement, so the NPL took on the role of making sure that British physics could readily be made available for the service of empire.

Britain's physicists duly responded when war was declared in 1914. Cavendish-trained graduates flocked to put their physics to work helping out with the war effort. Unsurprisingly, the National Physical Laboratory with Glazebrook still at its head was an important center for war work. Since 1909, Glazebrook along with Rayleigh had been at the core of the Advisory Committee for Aeronautics, established "for general advice on the scientific problems arising in connection with the work of the Admiralty and War Office in aerial construction and navigation."[21] Rayleigh's own earlier work on aerodynamics played an important role. When war broke out, the committee was important in bringing academic physicists into the fold. Physicists such as Frederick Lindemann and Frederick Aston found their way into aeronautics research at Farnborough and the Royal Aircraft Factory. Airplanes were not British physics' only contribution either. Cavendish men worked on explosives and on problems of submarine detection as well—inaugurating a tradition of scientific boffinry that lasted well into the twentieth century. The Great War in Britain was a physicists' war to an extent that is now often forgotten.

[21] R. J. Strutt, *Life of Lord Rayleigh* (London: Edward Arnold & Co., 1924), 338.

Under the aegis of the Admiralty's Board of Invention and Research labored the cream of high Victorian and Edwardian physicists, including J. J. Thomson, Oliver Lodge, Ernest Rutherford, and William Crookes.

Like the NPR in Britain, the Physikalisch-Technische Reichsanstalt was at the outset the fulcrum of German physicists' contributions to their country's war effort. When war broke out, Emil Warburg, the Reichsanstalt's president since 1905, immediately placed his institution at the service of the Ministry of War. This, after all, was one of the purposes for which the Reichsanstalt had been established—to put physics to work for the glory of the Reich. Unlike their British counterparts, however, who knew full well what to do with the National Physical Laboratory's resources, the German Imperial Army and Navy were not sure what use to put this help to. By 1915 half of the Reichsanstalt's personnel had either enlisted in the armed forces or been sent to carry out industrial war work. Those who remained in Berlin turned their resources to testing and improving war materiel. They carried out work on electrical equipment for the military, developing nonmagnetic alloys for the Navy and researching artillery ballistics. Soured by defeat, Warburg recalled later that during the war the physicists remaining at the Reichsanstalt "had been greatly burdened by testing work and by conducting largely unimportant work for the army."[22] Following Germany's surrender in 1918, the Reichsanstalt, product as it was of the short-lived Reich's aspirations for the new century, was a sad ghost of its old imperial self.

Physics was a combative business by the beginning of the twentieth century. Far more than just individual pride and amour propre were at stake in the proper allocation of credit for discovery, for example. National interest was at stake as well. Physics was an arena where countries as well as individuals competed for kudos. The lavish last farewell accorded Lord Kelvin was an indication of the important role physics had come to play in national self-images. The same can be said of the eulogies for Helmholtz in Germany, or a little earlier in the nineteenth century the impressive funeral in Washington for Joseph Henry, the Smithsonian Institution's first secretary and the United States' premier man of science. The resources that physics could offer for national competitiveness were increasingly recognized and regarded as making the science a prize well worth fighting for. Battles over international physical standards had real consequences in terms of control over increasingly crucial and strategic industries as well. When the long anticipated war eventually broke out in

[22]D. Cahan, *An Institute for an Empire*, 226.

1914, physics was ready to do its bit. That was the downside of physicists' success in making their discipline an important arm of early twentieth-century nation-states' apparatus.

## The End of the Ether

The ether was the crowning glory of late nineteenth-century physics. It was a massive success story that seemed to its promoters to be as real as the telegraph cables and electrical paraphernalia whose behavior it was originally invoked to explain. For physicists such as Oliver Lodge and William Thomson, the ether simply was not something about whose existence there could be any really serious doubt. They had very good reasons for this confidence as well. The ether was a formidable and powerful explanatory tool. The all-pervasive medium contained the mechanism through which the grand doctrine of the conservation of energy operated and became manifest. It explained the operations of telegraphs and telephones; it was the medium by which electromagnetic waves traveled through otherwise empty-seeming space; it transmitted messages from the distant stars to physicists' laboratories on Earth. Late nineteenth-century physicists were well versed in the ether's idiosyncrasies. They had developed powerful mathematical and experimental technologies to unravel its secrets. They were as familiar with it as they were with any of the other tools of the physicist's trade. Within a couple of decades of the century's end, however, the ether was dead. It had been revealed as a baroque fantasy better fitted for the amused condescension of a new generation than for any serious consideration on their part.

When J. J. Thomson announced in 1897 his demonstration that cathode rays were composed of a stream of negatively charged particles or corpuscles—later renamed electrons (a word coined by G. F. FitzGerald's physicist uncle, George Johnstone Stoney, and borrowed by Joseph Larmor, a prominent ether physicist, three years earlier)—his findings were in no way interpreted as an attack on the ether. On the contrary, most British physicists were already agreed that the rays were composed of particles of some sort. In any case, in Thomson's own view, his subatomic corpuscles were really vortices in the ether. What Thomson's work did suggest, however, was that the traditional notion of atoms as the smallest units of matter needed rethinking. These newfangled corpuscles were smaller than the smallest known atoms—Thomson's experiments deflecting the rays in an electrostatic field indicated that the corpuscles had one one-thousandth the mass of a hydrogen atom. The experimental technologies that Thomson deployed in his experiments provided powerful new tools for further research as well. Ether theorists such as G. F. FitzGerald were unhappy with Thomson's notion that his corpuscles were constituents of atoms. That was why they suggested they were "electrons" instead. In ether terms, electrons were pure, nonmaterial packets of electricity in the ether. FitzGerald recognized the power of Thomson's experiments, though. If Thomson were right, then it "would be the beginning of great advances in science, and the results it would be likely to lead to in the near future might easily eclipse most of the other great discoveries of the nineteenth century."[23]

To his immediate contemporaries, Thomson's assertion that cathode rays were made up of particles that were themselves constituent parts of atoms seemed at best an intriguing hypothesis. Agreeing with FitzGerald, they needed more data. As the *Electrician* editorialized: "Prof. J. J. Thomson's explanation of certain cathode ray phenomena by the assumption of the divisibility of the chemical atom leads to so many transcendentally important and interesting conclusions that one cannot but wish to see the hypothesis verified at an early date by some crucial experiment."[24] It was not too long, however, before such subatomic particles proliferated. Thomson's notion of subatomic particles seemed a promising way of accounting for the new phenomenon of radioactivity just being discovered by the Curies. Some of those emanations could be studied experimentally in a similar fashion and shown to consist of streams of particles of definite mass and velocity. In the early 1900s Thomson was speculating about the structure of the atom, regarding it as "built up of a number of corpuscles in equilibrium or steady motion under their mutual repulsions and a central attraction: it is surprising what a lot of interesting results come out."[25] He was not the only one engaged in such speculations. The Cavendish-trained Charles T. R. Wilson's invention of the cloud chamber in 1911 provided a powerful new tool for the investigation of these strange particles. Experimenters could now see the tracks left behind by individual corpuscles.

Leaving his negatively charged cathode ray corpuscles behind him, J. J. Thomson was working during the 1900s on what he called "positive rays" from gas discharge tubes. He was looking for positively charged analogues of his corpuscles. From 1910 onwards with Frederick Aston

---

[23] Quoted in I. Falconer, "Corpuscles, Electrons, and Cathode Rays," 273.

[24] Quoted ibid., 274.

[25] Quoted in I. Falconer, "J. J. Thomson's Work on Positive Rays," 267.

as his research assistant he started developing new and more powerful ways of studying and making visible the basic constituents (as he saw them) of matter. As with Wilson's cloud chamber, these were technologies that left tangible—and permanent—traces of the movements of otherwise invisible entities. They provided graphic evidence that there was more to this universe of proliferating subatomic particles than a theorist's reverie. To the experimenters who worked with them, subatomic particles seemed material enough. They could be manipulated and directed at will. The bright patches on photographic plates and fluorescent screens were proof of their passage. They provided grist as well for new models of atomic structure that themselves raised new problems. If, as Thomson's protégé Ernest Rutherford argued, atoms were made up of electrons orbiting around a central nucleus for example, why were they stable? According to Maxwell's theory, the orbiting electrons should emit electromagnetic waves. As a result they should collapse in a fraction of a second as their orbital energy disappeared. Not to do so without another source of energy was an apparent contradiction of the inexorable and sacred law of the conservation of energy.

In Germany, the new breed of theoretical physicists were worrying away at problems to do with the theory of radiation. Max Planck in Berlin was particularly interested in trying to integrate thermodynamics and electromagnetism. Planck had been extraordinary professor of physics at Berlin since 1889 and following his mentor Helmholtz's death was increasingly regarded as the Prussian capital's preeminent theoretician. During the late 1890s he was looking for a mathematical expression that could describe the distribution of energy across the electromagnetic spectrum—an equation that linked energy with the frequency of electromagnetic waves. The same constant factor kept on turning up in his calculations. This magic number, dubbed Planck's constant (symbolized in equations as $h$), was interpreted by its inventor as the "quantum of action." The number was soon turning up elsewhere. In particular the young Danish physicist Niels Bohr used Planck's constant to solve Rutherford's problem of the stability of his atomic model. According to Bohr, atoms were made up of electrons orbiting around a central core in discrete orbits defined by Planck's constant. They could move from one orbit to another only by absorbing or emitting energy in exact packets, defined by Planck's constant. This was peculiar. It was starting to look as if the waves of energy traveling through the ether traveled in discrete chunks—just as if they were made up of particles. According to Bohr, energy traveled in quanta.

While some turn-of-the-century physicists were messing around with subatomic particles—effectively in their view investigating the small-scale structure of the ether—others were investigating the ether's properties on a grander scale. In 1887 two American experimental physicists, Albert Abraham Michelson and Edward Morley, had come up with an intriguing observation. They had been carrying out an experiment to try to detect the movement of the Earth through the ether and had come up with a null result—they could not find the ether drift they were searching for. It looked as if, according to the Michelson-Morley experiment at least, the ether was somehow dragged along by the Earth's surface as the planet orbited the Sun. This was a peculiar—and rather irritating—conclusion since other aspects of ether theory assumed that the ether was not convected in this way. Oliver Lodge carried out a famous experiment with his enormous whirling machine to try to reproduce this dragging effect artificially, but to no avail. One suggestion that made sense of the Michelson-Morley observation and reconciled it with the rest of physics was the hypothesis offered in 1889 by G. F. FitzGerald, who pointed out that if matter moving through the ether contracted in the direction of its motion, the null result was precisely what was to be expected.

Few paid attention at first to FitzGerald's bizarre suggestion. During the late 1890s, however, Joseph Larmor, probably Britain's premier ether physicist, was working out his own novel speculations and by 1897 had calculated that the Michelson-Morley null result was actually a requirement of his new theory. It was also implied by this new theory that electrical systems moving through the ether would contract in exactly the manner predicted by FitzGerald. The Dutch physicist Hendrik Antoon Lorentz had came to a similar conclusion in 1892 on the basis of his theoretical work on high-velocity electrons and developed a set of mathematical transformations (later to be known as the Lorentz-FitzGerald transformations) describing the contraction. Lorentz had applied Maxwell's equations on a microscopic scale, showing that the contraction was the product of the way the electromagnetic forces between his electrons interacted with the stationary ether. The hunt was soon on among experimenters to find ways of experimentally detecting this peculiar phenomenon, particularly in view of ominous claims that Michelson and Morley might have got their experiment wrong and that there was indeed a measurable motion through the ether. By now an important plank of ether physics depended on there being no such thing. Frederick Trouton, following a suggestion by FitzGerald before his death, argued that there might be an electrical way of testing the Michelson-Morley experiment. Larmor reckoned that he could show from this experiment that anything other than a null result would be tantamount to the heresy of a perpetual motion machine. Trouton fell out with his theoretical masters, however, by continuing to try to measure the effects of the Earth's movement through the ether—he was intrigued by the possibility raised by Larmor that inexhaustible energy might be mined from the ether in this fashion.

Observers of Trouton's disagreements with his erstwhile mentors commented that he was going against received wisdom. The possibility of measuring the Earth's movement through the ether was "denied by one school of physicists among whom must be placed Larmor, Einstein, and . . . Sir O. Lodge."[26] That was how Einstein's announcement of his special theory of relativity in 1905 was first read in Britain. It was Einstein's work, however, that would over the following decade make the ether start coming apart at the seams. Albert Einstein in 1905 was a humble patent examiner in Switzerland, having graduated from Zurich Polytechnic a few years previously. In his soon-to-be-revolutionary paper "On the Electrodynamics of Moving Bodies," published in the *Annalen der Physik*, Einstein introduced two new principles into physics that led eventually to a completely new understanding of the nature of space and time. According to his principle of relativity there was no privileged, absolute perspective from which to view events in the universe. Everything was relative—except for the velocity of light, which remained the same in all frames of reference. This was the second principle—the constancy in all frames of the velocity of light. There was no such thing as Newtonian absolute space and therefore, of course, no ether. The ether as a hypothesis was superfluous to requirements. Einstein might have been dealing with the same kinds of problems in the electrodynamics of moving bodies as was Larmor, but he was coming at them from a very different perspective. His work was informed by the tradition of German theoretical physics rather than Cambridge ether physics.

Einstein's little publication helped to rescue him from comparative obscurity in his Zurich patent office and soon catapulted him onto the world stage. On the way he struck the death knell of ether physics. Many British physicists still committed to the ether found it easy at first to accommodate Einstein's rather outré ideas on relativity into their worldview. To a nonexpert observer at least it was not easy to see there was any great difference between what Einstein had to say and Larmor's latest

---

[26]Quoted in A. Warwick, "The Sturdy Protestants of Science," 325.

theories. For a decade after 1905 Einstein—now a fully accredited member of Germany's academic theoretical physics community—beavered away at the problem of generalizing his new theory. The trick was to find a way of accounting for gravity. He produced his coup in 1916 with his "Die Grundlage der allgemeinen Relativitätstheorie," again published in the prestigious *Annalen der Physik*. Here he argued that gravity could be understood as the curvature of space-time—a geometrical property of non-Euclidean space. He applied his theory to solve the problem of the planet Mercury's hitherto anomalous orbit around the sun. There was another test as well. According to Einstein's theory, light passing by a body with a strong gravitational field—like the Sun—should bend. The announcement by Einstein's British fan, Arthur Eddington, in 1919 that he had observed just that during that year's total eclipse of the Sun, was regarded by many as the final nail in the ether's coffin.

The end of the ether as a viable physical construct marked the end of the nineteenth century's imperial physics as well. The ether had encapsulated the hopes and the hubris of late Victorian physical science. Working out and minutely measuring its intimate properties was going to herald the end of physics—the final theory of everything. Its proponents were confident that the ether was as real as anything else in their physics. They could manipulate it as easily and as purposefully as any of the increasingly complex pieces of apparatus that littered their laboratories. It was the machine culture of late nineteenth-century Europe and America inscribed on the Universe. The ether looked plausible to its makers just because its ingredients were so familiar. The wheels and pulleys of the factory floor were transformed into celestial mechanics—a process that had the virtue of ennobling the one while giving the other a much-needed practical edge. Just as the ether was coming apart at the seams during the early years of the twentieth century, so was the culture that produced it. It had been the product of a generation of physicists who were on the whole supremely confident and self-assured of the progressive nature of their science and its potential for society. Successive generations were to have far fewer illusions.

## Conclusion

From a vantage point over a century later, much of nineteenth-century physics looks a little peculiar, if not downright bizarre. Late Victorian models of the ether in particular look odd, with their gears and pulleys and wheels. It is far easier to sympathize with the French physicist and philosopher Pierre Duhem and his ridicule of such constructions than it is to understand the attraction they held for their supporters. Most misplaced to modern eyes is the immense confidence that these late nineteenth-century theorists had in their science. We know, after all, that within less than a generation those fantastical constructions were to come crashing down about their ears. It is still tempting for many historians of science and scientists to try to overlook what appear from today's perspective to be the errors and excesses of past scientific heroes. By obscuring or forgetting the bits that seem odd, incongruous, or just out of place in sober-minded science we hope that we can continue to present the history of physics (and that of other sciences) as a grand, triumphant narrative of truth emerging out of ignorance. In that at least we still have a great deal in common with our Victorian predecessors and their faith in progress. Leaving out, or reducing to the level of amusing anecdote, those parts of nineteenth-century physics that fail to conform comfortably with our present-day views of what physics is—or should be—about is to do the past a great disservice. It is also a denigration of the full richness and complexity of contemporary scientific culture.

One feature of the history of nineteenth-century physics that may strike modern readers as particularly surprising is the comparatively recent date at which physics (or any of the other sciences for that matter) became an academic discipline in anything like the contemporary sense. Until well into the nineteenth century—in Britain at any rate—practicing physicists were as likely, if not more likely, to be found outside as inside universities. Michael Faraday, perhaps the greatest of nineteenth-century scientific heroes, never set foot in a university as either student or teacher. This is often portrayed as an achievement on Faraday's part—evidence of his innate genius or heroic efforts at self-education. It is, of course, nothing of the sort. On the contrary, Faraday's informal apprenticeship at the Royal Institution under Sir Humphry Davy's tutelage was probably as good as it got in terms of an early nineteenth-century education in chemistry and natural philosophy. Certainly for the first half of the nineteenth century, even in countries such as France and some of the German lands where natural philosophy or physics had a more prominent place in the university curriculum, teaching to produce physicists—as opposed to well-rounded individuals—simply was not part of what universities typically did. It was even more uncommon for universities to actively encourage, let alone financially support, any activity that might now be recognized as experimental research in physics. Making physics into a university discipline with attendant regimes of training, career structure, and opportunity if not obligation to do research was very much an achievement of the second half of the nineteenth century.

In many ways, nineteenth-century natural philosophers and physicists were a far more varied bunch than their modern counterparts. They had not undergone similar training regimes, neither were they obliged to acquire particular qualifications to become accredited practitioners of their science. Even where university teaching in natural philosophy existed, it was not designed with the aim of providing accreditation or training for budding physicists. The argument for making natural philosophy part of the university curriculum was that physics could be a valuable means of inculcating a liberal education—of producing well-rounded, well-informed potential members of the ruling classes. Most hopeful nineteenth-century physicists then had to make their careers up as they went along. There simply was no obvious, well-worn route for them to follow. The lucky ones were independently wealthy, vocationally minded gentlemen such as Charles Babbage or John Herschel who could pursue their interests without undue concern for the need to make a living. Those less fortunate had to find ways of earning a crust that allowed them to practice their science as well. The result was that for most of the period, the question of what kind of person the physicist should be—how trained, how employed, how situated in society—was very much open to different interpretations. Gentlemen of science in the earlier part of the century, for example, often looked down their noses at those who tried to profit from their science by taking out patents for inventions.

At least partly as a result of the socially insecure place of physics and physicists within nineteenth-century culture, cultivating the public mattered for practitioners then to a degree that may seem surprising now. Public lectures for a popular audience were a significant source of income for many men of science. Some performers were truly popular as well. Fashionable London flocked in its hundreds to Michael Faraday's lectures at the Royal Institution. John Tyndall later in the century was a popular draw as well. The real superstars such as Lord Kelvin even went on international tours. Audiences crammed into galleries of practical science, sold out the Great Exhibition of 1851 and rushed to international expositions throughout Europe and America, anxious to see and admire the latest products of science and industry. They did so because to some degree at least they shared in and endorsed (though not entirely uncritically) the ethos of optimistic progress through science that those enterprises represented. Polite, middle-class, and working-class society in different ways accepted physics as part of their culture. Knowing physics was part of what it meant to be a cultivated person in nineteenth-century Europe and America. As well as the ever-present exhibitions and lectures, newspapers and popular magazines routinely reported the latest discoveries and pronouncements. Popular books and pamphlets explained the latest inventions and theories to their readers at levels often far more technical than would be acceptable in today's mass media.

Because of this popularity—and because of the assumptions of its practitioners and their audiences concerning what kind of role science should play in society—nineteenth-century physics could be a political hot potato. The nebular hypothesis is probably the best example of this. Lessons drawn from the heavens concerning the progressive unfolding of natural law were quite straightforwardly accepted as having important consequences for the way in which society should be organized. Natural philosophical systems that veered too close to the materialistic were castigated by conservative churchmen and politicians for undermining the social fabric and giving the hoi polloi ideas above their station. Radicals just as cheerfully turned to physics for ammunition. Karl Marx and Friedrich Engels lauded William Robert Grove's doctrine of the correlation of physical forces and the theory of the conservation of energy as politically correct (that is to say materialist and progressive) science. Access to physics could be a political issue too. Radical spokespersons early in the century attacked the gentlemen of science for trying to maintain a monopoly on knowledge—for insisting that only they and their kind could be trusted to do science. Their radical opponents argued that on the contrary science should be available to all and that physics in particular was the proper province of those who worked for a living and therefore had hands-on knowledge of machines and of natural processes.

By their own lights, by the end of the nineteenth century physicists and their commentators reckoned their science as a success story. They marveled at physics' massive strides forward in knowledge over the last century. They celebrated their physics as tangible evidence of their culture's progressive modernity. Not everyone was a fan, of course. The overarching problem was the materialism of modern science as a whole. Antivivisectionists attacking physiology for its penchant for animal experimentation held that this was only the thin edge of the wedge. More than a few physicists evinced at least some sympathy with this argument. Several eminent late Victorian physicists looked sympathetically, if skeptically, at psychics and spiritualists. They thought that the world of the spirit could be embraced within the ambit of physics as well. On the whole, though, physicists saw the triumph of their discipline inscribed on the world around them. It was visible in the factories and in the ever-growing telegraph network. By the beginning of the twentieth century it was there in the cinema, the motor car, and the radio; it even powered flight. The great thing about their physics for the Victorians was that it really seemed to work. It was a science that delivered the goods.

While most late nineteenth-century physicists were supremely confi-dent of the security of the content of their science—that they had at least a general outline of a comprehensive account of the universe—they were rather less confident concerning the security of the institutions on which their science depended. To many of them, their hard-fought-for institu-tions and the cultural place of physics that they had worked so diligently to establish still seemed very fragile. Turn-of-the-century physicists were keenly aware that their science depended on their institutions and that those institutions' continuing vitality depended on physics' being able to maintain its privileged cultural position in nineteenth-century society as the ultimate authority on nature. Senior members of the profession—and physics was by now a profession—had lived through and participated in the struggles that had been needed to establish physics in its current po-sition of cultural authority. That this privileged social place might yet slip from their grasp was a real concern. This seems particularly ironic from a modern perspective, of course. It is after all the institutions that late nineteenth-century physicists labored to establish that have survived to the present day largely unchanged while large parts (though by no means all) of their physics has fallen by the wayside. Social structures turned out in the end to be considerably more durable than physical theories.

This book has, I hope, given some sense of how physics fitted into nineteenth-century culture. By doing so, it should also have instilled some sense of the historically contingent nature of modern science and its institutions. Science is not a given. It is a cultural achievement of immense and unprecedented significance. A hundred odd years on since the scientific struggles related in this book, physics is part of the everyday fabric of our lives. This sentence is being written with a machine that could have been built only with the detailed knowledge of the workings of subatomic particles that modern physics provides. As this book shows, that knowledge did not come from nowhere. Since physics is a product of culture, as the nineteenth century recognized, it is also part of a common culture. The shape of modern scientific institutions, the status of scientific experts, their relationship to government and to industry are not engraved in stone. It is up to citizens of the twenty-first century to decide whether and how they value physics and physicists—what role they will play in this century's culture. Nineteenth-century physicists' awareness of the fragility of their fledgling discipline and the ease with which the cultural niche they had carved for themselves could be whittled away meant that they worked hard to engage their publics. That is one lesson at least that the history of imperial physics has to offer the postmodern age.

The first volume of Edmund Whittaker, *A History of the Theories of Aether and Electricity* (London: Thomas Nelson, 1951), is still worth consulting as a general survey of nineteenth-century physics, though readers should beware the anachronistic mathematical notation and, if they carry on into the second volume, the idiosyncratic refusal to credit Einstein with the theory of relativity. Robert Purrington, *Physics in the Nineteenth Century* (New Brunswick: Rutgers University Press, 1997), is a good recent survey. Another perspective can be found in Mary Jo Nye, *Before Big Science* (Cambridge: Harvard University Press, 1996). Peter Harman, *Energy, Force, and Matter* (Cambridge: Cambridge University Press, 1982), surveys the rise of energy physics.

### Chapter 2. A Revolutionary Science

The classic account of French physics under Laplace is Robert Fox, "The Rise and Fall of Laplacian Physics," *Historical Studies in the Physical Sciences*, 1974, 4: 89–136. Eugene Frankel, "Corpuscular Optics and the Wave Theory of Light: The Science and Politics of a Revolution in Physics," *Social Studies of Science*, 1976, 6: 141–84, fleshes out some of the details with regard to the key science of optics. The institutional background and political networks are surveyed in Maurice Crosland, *The Society of Arcueil* (London: Heinemann, 1967), and *Science under Control* (Cambridge: Cambridge University Press, 1992). For a comparison of French and English mathematical traditions see Joan Richards, "Rigor and Clarity: Foundations of Mathematics in France and England, 1800–1840," *Science in Context*, 1991, 4: 297–319. The background to mathematics in Cambridge is discussed in John Gascoigne, *Cambridge in the Age of Enlightenment: Science, Religion, and Politics from the Restoration to the French Revolution* (Cambridge: Cambridge University Press, 1989). For the "analytical revolution" see Harvey Becher, "Radicals, Whigs, and Conservatives: The Middle and Lower Classes in the Analytical Revolution in Cambridge in the Age of Aristocracy," *British Journal for the History of Science*, 1995, 28: 405–26, and William Ashworth, "Memory, Efficiency, and Symbolic Analysis: Charles Babbage, John Herschel, and the Industrial Mind," *Isis*, 1996, 87: 629–53. Simon Schaffer, "Babbage's Intelligence: Calculating Engines and the Factory System," *Critical Inquiry*, 1994, 21: 203–27, expands on Babbage's project. Harvey Becher, "William Whewell and Cambridge Mathematics," *Studies in History and Philosophy of Science*, 1980, 11: 1–48, takes the Cambridge story further. The rise of the mathematics Tripos and the coaching system is discussed in Andrew Warwick, "Exercising the Student Body: Mathematics and Athleticism in Victorian Cambridge," Christopher Lawrence and Steven Shapin (eds.), *Science Incarnate* (Chicago: University of Chicago Press, 1998), and "A Mathematical World on Paper: Written Examinations in Early Nineteenth-Century Cambridge," *Studies in History and Philosophy of Physical Science*, 1998, 29b: 295–319. More detail is provided in Andrew Warwick, *Masters of Theory: Cambridge and the Rise of Mathematical Physics* (Chicago: University of Chicago Press, 2003). Peter Harman (ed.), *Wranglers and Physicists* (Manchester: Manchester University Press, 1985), surveys the field. Joan Richards, *Mathematical Visions: The Pursuit of Geometry in Victorian England* (Boston: Academic Press, 1988), gives an overview of English mathematics. Christa Jungnickel and Russell McCormmach, *The Intellectual Mastery of Nature: Theoretical Physics from Ohm to Einstein* (Chicago: University of Chicago Press, 1986), provides an exhaustive survey of the institutional rise of German physics. David Cahan (ed.), *Herman von Helmholtz and the Foundation of Nineteenth-Century Science* (Berkeley: University of California Press, 1994), fleshes out the picture, as does Kathryn Olesko, *Physics as a Calling* (Ithaca: Cornell University Press, 1991).

## Chapter 3. The Romance of Nature

The impact of the Romantic movement on natural philosophy is surveyed in Andrew Cunningham and Nicholas Jardine (eds.), *Romanticism and the Sciences* (Cambridge: Cambridge University Press, 1990), especially D. Sepper, "Goethe, Colour, and the Science of Seeing"; W. Wetzels, "Johann Wilhelm Ritter: Romantic Physics in Germany"; and C. Lawrence, "The Power and the Glory: Humphry Davy and Romanticism." H. A. M. Snelders, "Romanticism and Naturphilosophie and the Inorganic Natural Sciences," *Studies in Romanticism*, 1970, 9: 193–215, and W. Wetzels, "Aspects of Natural Science in German Romanticism," *Studies in Romanticism*, 1971, 10: 44–59, deal with physics more particularly. See also Myles Jackson, "A Spectrum of Belief: Goethe's 'Republic' versus Newton's 'Despotism,'" *Social Studies of Science*, 1994, 24: 673–701, which deals with Goethe's anti-Newtonianism. Coleridge's romanticism and his views of natural philosophy are canvassed in Trevor Levere, *Poetry Realized in Nature: Samuel Taylor Coleridge and Early Nineteenth-Century Science* (Cambridge: Cambridge University Press, 1981). Davy's romanticism is discussed in David Knight, *Humphry Davy: Science and Power* (Cambridge: Cambridge University Press, 1996). The classic overview of conversion experiments and their role in the development of the theory of the conservation of energy is Thomas Kuhn, "Energy Conservation as an Example of Simultaneous Discovery," *The Essential Tension* (Chicago: University of Chicago Press, 1977). The origins of the Royal Institution and its politics of urbane utilitarianism are discussed in Morris Berman, *Social Change and Scientific Organization* (London: Heinemann, 1978). Conservation, conversion, and correlation are analysed in Peter Heimann, "Conversion of Forces and the Conservation of Energy," *Centaurus*, 1974, 18: 147–61; Geoffrey Cantor, "William Robert Grove, the Correlation of Forces, and the Conservation of Energy," *Centaurus*, 1976, 19: 273–90; Iwan Rhys Morus, "Correlation and Control: William Robert Grove and the Construction of a New Philosophy of Scientific Reform," *Studies in History and Philosophy of Science*, 1991, 21: 589–621. Faraday's contribution is discussed in David Gooding, "Final Steps to the Field Theory," *Historical Studies in the Physical Sciences*, 1981, 11: 492–505. Daniel Siegel, *Innovation in Maxwell's Electromagnetic Theory* (Cambridge: Cambridge University Press, 1991), and Peter Harman, *The Natural Philosophy of James Clerk Maxwell* (Cambridge: Cambridge University Press, 1998), discuss Maxwell; Bruce Hunt, *The Maxwellians* (Ithaca: Cornell University Press, 1991), gives an incisive analysis of his followers. Geoffrey Cantor and John Hodge (eds.), *Conceptions of Ether* (Cambridge: Cambridge University Press, 1981), surveys the consolidation of the ether.

## Chapter 4. The Science of Showmanship

The Galvani-Volta debate and its consequences form the focus for Marcello Pera, *The Ambiguous Frog* (Princeton: Princeton University Press, 1992). Volta's impact in revolutionary France is outlined in Geoffrey Sutton, "The Politics of Science in Early

Napoleonic France: the Case of the Voltaic Pile," *Historical Studies in the Physical Sciences*, 1981, **11**: 329–66. Faraday's massive contribution is laid out in L. Pearce Williams, *Michael Faraday* (London: Chapman and Hall, 1965), David Gooding and Frank James (eds.), *Faraday Rediscovered* (London: Macmillan, 1985), and Iwan Rhys Morus, *Michael Faraday and the Electrical Century* (London: Icon Books, 2004). James R. Hoffmann, *André-Marie Ampère* (Cambridge: Cambridge University Press, 1996), outlines Ampère's contribution. The importance of experiment and demonstration is established in David Gooding, "Experiment and Concept-Formation in Electromagnetic Science and Technology in England," *History and Technology*, 1985, **2**: 229–44; Iwan Rhys Morus, "Different Experimental Lives: Michael Faraday and William Sturgeon," *History of Science*, 1992, **30**: 1–28; Iwan Rhys Morus, "Currents from the Underworld: Electricity and the Technology of Display in Early Victorian London," *Isis*, 1993, **84**: 50–69; and Iwan Rhys Morus, *Frankenstein's Children* (Princeton: Princeton University Press, 1998). Albert Moyer, *Joseph Henry* (Washington: Smithsonian Institution Press, 1997), discusses the American perspective. Brian Gee, "The Early Development of the Magneto-Electric Machine," *Annals of Science*, 1993, **50**: 101–33, looks at experiment and utility, while Iwan Rhys Morus, "Manufacturing Nature: Science, Technology, and Victorian Consumer Culture," *British Journal for the History of Science*, 1996, **29**: 403–34, relates it to exhibition. For electric light see Wolfgang Schivelbusch, *Disenchanted Night: The Industrialization of Light in the Nineteenth Century* (Berkeley: University of California Press, 1988). Jeffrey Kieve, *The Electric Telegraph* (Newton Abbott: David and Charles, 1973), gives a social history of telegraphy; Bruce Hunt, "Michael Faraday, Cable Telegraphy, and the Rise of Field Theory," *History of Technology*, 1991, **13**: 1–19, examines its impact on later physical theory. Yakub Bektas, "The Sultan's Messenger: Cultural Constructions of Ottoman Telegraphy, 1847–1880," *Technology and Culture*, 2000, **41**: 669–96, provides a cross-cultural perspective. Ken Beauchamp, *Exhibiting Electricity* (London: Institute of Electrical Engineers Press, 1997), and Carolyn Marvin, *When Old Technologies Were New* (Oxford: Oxford University Press, 1988), look at electrical exhibitions later in the century. Robert Brain, *Going to the Fair: Readings in the Culture of Nineteenth-Century Exhibitions* (Cambridge: Whipple Library, 1993), provides some interesting perspectives.

## Chapter 5. The Science of Work

The context of engineering in revolutionary and pre-Revolutionary France is laid out in Ken Adler, *Engineering the Revolution* (Princeton: Princeton University Press, 1997). The introduction to Sadi Carnot, *Reflections on the Motive Power of Fire* (Manchester: Manchester University Press, 1986; originally published 1824), translated and edited by Robert Fox, lays out the context for Carnot, as well as being an invaluable edition of his writings. Another, earlier translation, Eric Mendoza (ed.), *Reflections on the Motive Power of Fire and other Papers on the Second Law of Thermodynamics by E. Clapeyron and R. Clausius* (New York: Dover Publications, 1960), also contains translations of key papers by Clapeyron and Clausius. Donald Cardwell, *From Watt to Clausius* (Ames: Iowa State University Press, 1989), gives the steam background and context. Robert Fox, *The Caloric Theory of Gases from Lavoisier to Regnault* (Oxford: Oxford University Press, 1971), provides useful detail. Crosbie Smith, *The Science of Energy* (Chicago: University of Chicago Press, 1999), is a brilliant and indispensable account of English, Irish, and Scottish developments. More details on Lord Kelvin are found in Crosbie Smith and Norton Wise, *Energy and Empire: A Biographical Study of Lord Kelvin* (Cambridge: Cambridge University Press, 1989). Crosbie Smith and Norton Wise, "Measurement, Work, and Industry in Lord Kelvin's Britain," *Historical Studies in the Physical Sciences*, 1986, **17**: 147–73; Norton Wise (with the collaboration of Crosbie Smith), "Work and Waste: Political Economy and Natural Philosophy in Nineteenth-Century Britain," *History of Science*, 1989, **27**: 263–301, 391–449; 1990, **28**: 221–61; and Norton Wise, "Mediating Machines," *Science in Context*, 1988, **2**: 77–113, are all well worth reading. Joule's contributions are laid out in Donald Cardwell, *James Joule* (Manchester: Manchester University Press, 1989), and the details of the paddlewheel experiment are the focus of Heinz Otto Sibum, "Reworking the Mechanical Equivalent of Heat: Instruments of Precision and Gestures of Accuracy in Early Victorian England," *Studies in History and Philosophy of Science*, 1995, **26**: 73–106. The Manchester context more generally is outlined in Robert Kargon, *Science in Victorian Manchester: Enterprise and Expertise* (Manchester: Manchester University Press, 1977). Mayer's contribution is exhaustively covered in Ken Caneva, *Robert Mayer and the Conservation of Energy* (Princeton: Princeton University Press, 1993). Yehuda Elkana, *The Discovery of the Conservation of Energy* (London: Hutchinson, 1974), identifies Helmholtz as the discoverer. There is an interesting perspective on Helmholtz in Anson Rabinbach, *The Human Motor* (Berkeley: University of California Press, 1990). Some of his popular scientific writings are collected in David Cahan (ed.), *Science and Culture* (Chicago: University of Chicago Press, 1995). Joe Burchfield, *Lord Kelvin and the Age of the Earth* (Chicago: University of Chicago Press, 1975), lays out the disputes between thermodynamics and Darwinian evolutionary theory. The history of statistical mechanics is discussed in Stephen G. Brush, *The Kind of Motion We Call Heat* (Amsterdam: North Holland, 1976), and *Statistical Physics and the Atomic Theory of Matter from Boyle and Newton to Landau and Onsanger* (Princeton: Princeton University Press, 1983). Ted Porter, *The Rise of Statistical Thinking, 1820–1900* (Princeton: Princeton University Press, 1986), provides important context. Some of the cultural dynamics are discussed in Greg Myers, "Nineteenth-Century Popularizations of Thermodynamics and the Rhetoric of Social Prophecy," Patrick Brantlinger (ed.), *Energy and Entropy* (Bloomington: Indiana University Press, 1989).

### Chapter 6. Mysterious Fluids and Forces

Some of the background to Victorian developments in discharge tube physics is outlined in Frank James, "The Study of Spark Spectra, 1835–1859," *Ambix*, 1983, **30**: 137–62. Crookes's work is the focus for Robert deKosky, "Spectroscopy and the Elements in the Late Nineteenth Century," *British Journal for the History of Science*, 1973, **6**: 400–23, and "William Crookes and the Fourth State of Matter," *Isis*, 1976, **67**: 36–60. Crookes is also discussed in Frank James, "Of Medals and Muddles: The Context of the Discovery of Thallium: William Crookes's Early Spectro-Chemical Work," *Notes and Records of the Royal Society*, 1984, **39**: 65–90, and Hannah Gay, "Invisible Resource: William Crookes and His Circle of Support," *British Journal for the History of Science*, 1996, **29**: 311–36. Bruce Hunt, "Practice vs. Theory: The British Electrical Debate, 1888–1891," *Isis*, 1983,

74: 137–62, deals with the British Maxwellians' efforts to find electromagnetic waves. Hertz's discovery of radio is the focus of Jed Buchwald, *The Creation of Scientific Effects* (Chicago: University of Chicago Press, 1994). Hugh Aitken, *Syntony and Spark* (New York: Wiley, 1976), and Sungook Hong, "Marconi and the Maxwellians: The Origins of Wireless Telegraphy Revisited," *Technology and Culture*, 1994, 35: 717–49, deal with the aftermath. Alison Winter, *Mesmerized* (Chicago: University of Chicago Press, 1998), discusses some of the mesmeric background to psychic research. Janet Oppenheim, *The Other World* (Cambridge: Cambridge University Press, 1985), deals with spiritualism in general. Richard Noakes, "Telegraphy Is an Occult Art: Cromwell Fleetwood Varley and the Diffusion of Electricity to the Other World," *British Journal for the History of Science*, 1999, 32: 421–59, provides an important analysis of the close relationship between psychic and telegraphic research, as does his "Instruments to Lay Hold of Spirits: Technologizing the Bodies of Victorian Spiritualism," Iwan Rhys Morus (ed.), *Bodies/Machines* (Oxford: Berg Publishers, 2002). Arne Hessenbruch, "Calibration and Work in the X Ray Economy, 1896–1928," *Social Studies of Science*, 2000, 30: 397–420, discusses the consolidation of X-ray technologies. Mary Jo Nye, "N-Rays: An Episode in the History and Psychology of Science," *Historical Studies in the Physical Sciences*, 1980, 11: 125–56, gives the details of the "discovery" of N-rays. Sorya Boudia, "The Curie Laboratory: Radioactivity and Metrology," *History and Technology*, 1997, 13: 249–65, and Xavier Roqué, "Marie Curie and the Radium Industry: A Preliminary Sketch," *History and Technology*, 1997, 13: 267–91, deal with the origins and early development of radioactivity. A recent and informative biography is Susan Quinn, *Marie Curie: A Life* (New York NY: Simon and Schuster, 1995). A more popular view of radioactivity is Catherine Caufield, *Multiple Exposures: Chronicles of the Radiation Age* (Harmondsworth: Penguin Books, 1989).

### Chapter 7. Mapping the Heavens

Robert Smith, "A National Observatory Transformed: Greenwich in the Nineteenth Century," *Journal of the History of Astronomy*, 1991, 22: 5–20, provides a good overview of industrial astronomy at the Greenwich Observatory. There is more detail on George Bidell Airy in Allen Chapman, "Science and the Public Good: George Bidell Airy and the Concept of a Scientific Civil Servant," Nicolaas Rupke (ed.), *Science, Politics, and the Public Good* (London: Macmillan, 1988). Airy and the discovery of Neptune are discussed in Allen Chapman, "Private Research and Public Duty: George Bidell Airy and the Search for Neptune," *Journal of the History of Astronomy*, 1988, 19: 121–39, and Robert Smith, "The Cambridge Network in Action: the Discovery of Neptune," *Isis*, 1989, 80: 395–422. Astronomical discipline is discussed in Simon Schaffer, "Astronomers Mark Time: Discipline and the Personal Equation," *Science in Context*, 1988, 2: 115–45, and Will Ashworth, "The Calculating Eye: Baily, Herschel, Babbage, and the Business of Astronomy," *British Journal for the History of Astronomy*, 1994, 27: 409–41. The invention of Greenwich time is discussed in Derek Howse, *Greenwich Time and the Longitude* (London: Philip Wilson Publishers, 1997), and in Iwan Rhys Morus, "The Nervous System of Britain: Space, Time, and the Electric Telegraph in the Victorian Age," *British Journal for the History of Science*, 2000, 33: 455–75. American innovations in the astronomical uses of time are discussed in Ian Bartky, *Selling the True Time*

(Stanford: Stanford University Press, 2000). Simon Schaffer, "The Nebular Hypothesis and the Science of Progress," James R. Moore (ed.), *History, Humanity, and Evolution* (Cambridge: Cambridge University Press, 1989), deals with the politics of the nebular hypothesis, while James Secord, "Behind the Veil: Robert Chambers and *Vestiges*," in the same collection, focuses on Chambers and his contribution to the debate. Simon Schaffer, "The Leviathan of Parsonstown: Literary Technology and Scientific Representation," Tim Lenoir (ed.), *Inscribing Science* (Stanford: Stanford University Press, 1998), deals with the impact of Lord Rosse's telescopic observations. The introduction of physics into astronomy is discussed in Holly Rothermel, "Images of the Sun: Warren de la Rue, George Bidell Airy, and Celestial Photography," *British Journal for the History of Science*, 1993, 26: 137–69, and in Alex Pang, "The Social Event of the Season: Solar Eclipse Expeditions and Victorian Culture," *Isis*, 1993, 84: 252–77, and "Victorian Observing Practices: Printing Technologies and Representations of the Solar Corona," *Journal of the History of Astronomy*, 1994, 25: 249–74; 1995, 26: 63–75. Stellar spectroscopy is discussed in Simon Schaffer, "Where Experiments End: Tabletop Trials in Victorian Astronomy," Jed Buchwald (ed.), *Scientific Practice* (Chicago: University of Chicago Press, 1995). The elder Herschel's views on lunar habitation are one of the topics of Simon Schaffer, "Herschel in Bedlam: Natural History and Stellar Astronomy," *British Journal for the History of Science*, 1980, 13: 211–39. The development of the extraterrestrial debate is followed in Michael Crowe, *The Extraterrestrial Life Debate* (Cambridge: Cambridge University Press, 1986).

## Chapter 8. Places of Precision

A comprehensive and recent survey of the rise of precision in physics is Norton Wise (ed.), *The Values of Precision* (Princeton: Princeton University Press, 1995). The rise of laboratory physics and regimes of measurement in Britain is discussed in Romualdas Sviedrys, "The Rise of Physics Laboratories in Britain," *Historical Studies in the Physical Sciences*, 1976, 7: 405–36. Particularly important are Graeme Gooday, "Precision Measurement and the Genesis of Physics Teaching Laboratories in Victorian Britain," *British Journal for the History of Science*, 1990, 23: 25–51, and "Teaching Telegraphy and Electrotechnics in the Physics Laboratory: William Ayrton and the Creation of an Academic Space for Electrical Engineering in Britain," *History of Technology*, 1991, 13: 73–111. Germany and the Physikalisch-Technische Reichsanstalt are dealt with in David Cahan, *An Institute for an Empire* (Cambridge: Cambridge University Press, 1989), "Werner Siemens and the Origins of the Physikalisch-Technische Reichsanstalt, 1872–1887," *Historical Studies in the Physical Sciences*, 1982, 12: 253–83, and "The Institutional Revolution in German Physics, 1865–1914," *Historical Studies in the Physical Sciences*, 1985, 15: 1–65. Cahan deals with German precision in "Kohlrausch and Electrolytic Conductivity: Instruments, Institutes, and Scientific Innovation," *Osiris*, 1989, 5: 167–85. An interesting German episode is discussed in Myles Jackson, *Spectrum of Belief* (Cambridge: MIT Press, 2000). French academic laboratories are discussed in Terry Shinn, "The French Science Faculty System, 1808–1914: Institutional Change and Research Potential in Mathematics and the Physical Sciences," *Historical Studies in the Physical Sciences*, 1979, 10: 271–332. Precision in U.S. laboratories is one theme of George Sweetnam, "Precision Implemented: Henry Rowland, the Concave Diffraction

Grating, and the Analysis of Light," N. Wise (ed.), *Values of Precision*. The introduction of physics into Japan is discussed in Yuzo Takahashi, "William Edward Ayrton at the Imperial College of Engineering in Tokyo: The First Professor of Electrical Engineering in the World," *IEEE Transactions on Education*, 1990, 33: 198–205, and in Graeme Gooday and Morris Low, "Technology Transfer and Cultural Exchange: Western Scientists and Engineers Encounter Late Tokugawa and Meiji Japan," *Osiris*, 1998, 13: 99–128. Bruce Hunt, "The Ohm Is Where the Art Is: British Telegraph Engineers and the Development of Electrical Standards," *Osiris*, 1994, 9: 48–63, deals with the importance of telegraphy for the standards industry, as does his "Doing Science in a Global Empire: Cable Telegraphy and Electrical Physics in Victorian Britain," Bernard Lightman (ed.), *Victorian Science in Context* (Chicago: University of Chicago Press, 1997). The imperial context of telegraphy is explored in Daniel Headrick, *The Invisible Weapon: Telecommunications and International Politics, 1851–1945* (Oxford: Oxford University Press, 1991). J. G. Crowther, *The Cavendish Laboratory* (New York: Science History Publications, 1974), gives the basic history of the laboratory. Romualdas Sviedrys, "The Rise of Physical Science at Victorian Cambridge," *Historical Studies in the Physical Sciences*, 1970, 2: 127–52, provides some more context concerning its origins. Simon Schaffer, "Late Victorian Metrology and Its Instrumentation: A Manufactory of Ohms," Robert Bud and Susan Cozzens (eds.), *Invisible Connections* (Bellingham: SPIE Press, 1992), deals brilliantly with the industrial ethos at Cambridge, with more of the same in his "Metrology, Metrication, and Victorian Values," Bernard Lightman (ed.), *Victorian Science in Context* (Chicago: University of Chicago Press, 1997). Women physicists at Cambridge are discussed in Paula Gould, "Women and the Culture of University Physics in Late Nineteenth-Century Cambridge," *British Journal for the History of Science*, 1997, 30: 127–49. The Cavendish under J. J. Thomson is the focus of Isobel Falconer, "J. J. Thomson and Cavendish Physics," Frank James (ed.), *The Development of the Laboratory* (London: Macmillan, 1989).

## Chapter 9. Imperial Physics

Most of the material in this last chapter is by way of summation of the argument developed in previous chapters. Thus, those interested in a closer look should look no further than the texts already mentioned. Some additional sources on particular issues are, however, worth mentioning. Physics and views of the body (particularly women's bodies), is discussed in Janet Oppenheim, *Shattered Nerves* (Oxford: Oxford University Press, 1991), and Cynthia Eagle Russett, *Sexual Science* (Cambridge: Harvard University Press, 1989). See also Iwan Rhys Morus, "A Grand and Universal Panacea: Death, Resurrection, and the Electric Chair," Iwan Rhys Morus (ed.), *Bodies/Machines* (Oxford: Berg Publishers, 2002). Views on late nineteenth-century science and the state are found in Robert Brain, *Going to the Fair: Readings in the Culture of Nineteenth-Century Exhibitions* (Cambridge: Whipple Library, 1993). Late nineteenth-century ether physics is discussed in Bruce Hunt, "Experimenting on the Ether: Oliver Lodge and the Great Whirling Machine," *Historical Studies in the Physical Sciences*, 1986, 16: 111–34, and "The Origins of the FitzGerald Contraction," *British Journal for the History of Science*, 1988, 21, 67–76. Oliver Darrigol, *Electrodynamics from Ampère to Einstein* (Oxford: Oxford University Press, 2000), provides a useful overview of theoretical developments.

Important new insights are to be found in Andrew Warwick, "On the Role of the FitzGerald-Lorentz Contraction Hypothesis in the Development of Joseph Larmor's Electronic Theory of Matter," *Archive for the History of the Exact Sciences*, 1991, **43**: 29–91; "Cambridge Mathematics and Cavendish Physics: Cunningham, Campbell, and Einstein's Relativity, 1905–1911," *Studies in History and Philosophy of Science*, 1992, **23**: 625–56; 1993, **24**: 1–25; "The Sturdy Protestants of Science: Larmor, Troughton, and the Earth's Motion through the Ether," Jed Buchwald (ed.), *Scientific Practice* (Chicago: University of Chicago Press, 1995). J. J. Thomson's discovery of the electron is outlined in Isobel Falconer, "Corpuscles, Electrons, and Cathode Rays: J. J. Thomson's 'Discovery' of the Electron," *British Journal for the History of Science*, 1987, **20**: 241–76; the aftermath is discussed in Isobel Falconer, "J. J. Thomson's Work on Positive Rays, 1906–1914," *Historical Studies in the Physical Sciences*, 1988, **18**: 265–310. A recent overview is Jed Z. Buchwald and Andrew Warwick (eds), *Histories of the Electron: The Birth of Microphysics* (Cambridge: MIT Press, 2001). The origins of Bohr's model of the atom are discussed in John Heilbron and Thomas Kuhn, "The Genesis of the Bohr Atom," *Historical Studies in the Physical Sciences*, 1969, **1**: 211–90. The literature on Einstein is massive. Abraham Pais, *Subtle Is the Lord* (Oxford: Clarendon Press, 1982), is still as good a place to start as any. For some indication of the British physics community's insecurities at the beginning of the last century and its role in the Great War, see Andrew Hull, "War of Words: The 'Public Science' of the British Scientific Community and the Origins of the Department of Scientific and Industrial Research, 1914–16," *British Journal for the History of Science*, 1999, **32**: 461–81.

# Index